SOLIDS LEVEL MEASUREMENT AND DETECTION HANDBOOK

SOLIDS LEVEL MEASUREMENT AND DETECTION HANDBOOK

JOE LEWIS

MOMENTUM PRESS, LLC, NEW YORK

Solids Level Measurement and Detection Handbook
Copyright © Momentum Press®, LLC, 2012.

First published by Momentum Press®, LLC
222 East 46th Street, New York, NY 10017
www.momentumpress.net

ISBN-13: 978-1-60650-254-9 (hard back, case bound)
ISBN-10: 1-60650-254-9 (hard back, case bound)
ISBN-13: 978-1-60650-256-3 (e-book)
ISBN-10: 1-60650-256-5 (e-book)

DOI: 10.5643/9781606502563

Cover design by Jonathan Pennell
Interior design by Exeter Premedia Services Private Ltd.,
Chennai, India

10 9 8 7 6 5 4 3 2 1

Printed in the United States of America

Dedication

To: Diana, my loving wife,

Don Morrison and Bob Becker, mentors early in my career,

Eric L. Lewis and Leonora R. Lewis, my parents,

Don Ginesi, flow measurement expert and best friend (after my wife).

Contents

Preface

By the end of 2008 I had reached the conclusion that the topic of solids level measurement and monitoring had been inadequately covered in academia and within instrumentation textbooks in general. A dedicated self-tutorial on the topic would be a huge service to a wide variety of people within a range of industries. The stage was set for this book to become reality, but I had yet to realize it.

I had already authored several papers and technical articles within the area of solids level measurement. The new business venture I had begun early in 2009 would continue my involvement in the area of solids level measurement and monitoring and I continued to explore new ways to provide users of solids level instrumentation the information and knowledge they would need to best choose and implement the myriad of products in this area, including the production of podcasts and blogs.

In the spring of 2009 the seed that would become this book was planted during one of many periodic lunches with my friend, automation expert, and magazine editor-in-chief Walt Boyes. By late 2009 a loose proposal for this book was being discussed and in the beginning of 2010 the idea had become a publishing contract with Momentum Press.

Now as I sit writing this Preface, I wonder at the outcome, the result. Will it be the self-tutorial I desire it to become? Will readers embark down the path of self-learning using this book as study guide, course work, and reference material? It is my goal for this to be the case, for each reader to walk away better off for having opened it, read sections, or the contents in its entirety.

I have designed each chapter to stand on its own or be read through in any order the reader chooses along with some or all of the other chapters. There is structure within each chapter, one at a time, together in a grouping, or in their entirety. So with this in mind I invite the reader to embark on a journey, one where you can proceed at your own speed and your own design. Study one chapter at a time or in a group. Test yourself. Come back to a chapter whenever you need to refresh your knowledge. It is my hope and desire for everyone to enjoy the learning from this text as much as I have enjoyed the research and preparation.

And who is the target reader? I have written this book to serve others, with the primary focus on those who find themselves in a variety of jobs including the process engineer, facility engineer, instrumentation technicians, maintenance specialists, managers of technical and process manufacturing disciplines, as well as those in school studying to move into any of these fields of endeavors where the processing, storage, or handling of powders and bulk solids will take place.

Introduction

It was 1964 and I was a young boy of 9 years. We were stationed at Ft. Dix in New Jersey. My dad was a career Army officer, which makes me an Army brat (emphasis on brat). I enjoyed the military family life and think it influenced me heavily, especially in my ability to pull up roots and move around frequently, at anytime and without notice.

During the summer my friends and I ran around the base and played "army," dug tunnels and generally got pretty dirty. We had lots of fun to say the least. One memory I have is hanging out in a wooded area overlooking a sand pit. Later I learned it was used as some sort of firing and exercise range, which explains finding empty cartridge shells and some ammunition containers (we collected these). We also found grenade pins and even fragments and components of grenades. We were lucky we never got caught or hurt.

The sand pit must have been in productive use at some point in time because there were two old rusty bins and other mechanical equipment in one area of the pit. Looking back it was probably the hulk of an old concrete batch plant or sand sorting machinery. We would climb on the bins and pretend it was some part of our great assault on a town during World War II. It's funny thinking back on it. Our guns, usually sticks, were Thompson submachine guns like the one Sgt. Saunders used in the TV series *Combat*, which ran from 1962 to 1967.

I remember one time when we were playing army at the sand pit, a friend of mine began throwing rocks at one of the old steel bins. He'd throw one up high, then another a little lower and then down toward the bottom. I had no idea what he was doing and thought it looked fun, so I started throwing rocks like that too. After a few throws I asked him what in the world we were doing this for. His reply stands out in my mind. "I'm trying to find out if there is anything in the bin and how full the bin is," he said. This was my first dealing with level detection and measurement of a bulk solid material. I guess God had a plan for me and kept me involved in the area of process bin level measurement throughout much of my adult life. Little did I know that I would end up writing a book on this subject.

For thousands of years man has been measuring the level of solid or liquid materials in a variety of vessels, including the ancient Egyptians who monitored the level of the Nile River with the goal of understanding the best times for irrigation. Knowing how much powder or granular material is in a bin or silo, and whether that material is present at some particular point such as a high or low point has been universally important throughout time. The need for measuring and monitoring material levels has changed very little. And the challenges in making these measurements still exist as well, even though technology has advanced and given us several solutions to choose from.

In reviewing other books on this or related subjects I found that only a chapter or two have been devoted to the subject of level measurement and monitoring, and those were focused on liquid level. There is often little more than a couple of paragraphs dedicated to the area of level measurement and monitoring of powder and bulk solid materials, if that. For many reasons the solids measurement area has had little attention, but the need has never been greater. Measuring the level of bulk solids presents numerous challenges not found in the liquid level measurement area.

The purpose of this book is to provide study materials on the subject of powder and bulk solids level measurement and detection and it is organized into several logical chapters. Each chapter is designed for study by the reader and will usually consist of the following parts: Chapter Objectives, Chapter Summary, Chapter Discussion, Study Questions, and Answers.

A word about how you should approach reading this book. It is not necessary to read it from cover to cover, but feel free if you like. It is designed and written so that the reader can review and study the material within any one or multiple chapters of interest. Flexibility to meet the needs of the reader is one of the objectives of this book.

Chapter Objectives: This defines what you should know or be able to do after completing your study of each chapter. There may be a single accomplishment or multiple points. The objectives will clarify what value should be added to your repertoire if we, together, have accomplished what we set out to do in the chapter.

Chapter Summary: Usually one or two paragraphs, the Chapter Summary boils down the Chapter Discussion. It acts as a framework or outline, if you will, to the essence of the chapter.

Chapter Discussion: The body of written work which discusses the subject matter of the chapter to accomplish the objectives. The Chapter Discussion may be broken into segments for clarification and organization.

Study Questions: These are review questions for the reader to use as a guide to study the chapter. Reviewing and studying the Chapter Discussion to answer each Study Question one at a time will help the reader "learn" the material presented and discussed in the chapter. I firmly believe in the study principle that states that we comprehend far more if we write down answers to questions. It is recommended, in order to get the most from this work and the material presented in each chapter, that you write down each question, research the answers within the Chapter Discussion, and write down the answers.

Answers: To assist you in your learning process we will include answers to the chapter Study Questions. Answers will be variable in length and may refer to sections within the chapter, other chapters within the book, or reference material. For the best learning experience it is suggested that you review the Chapter Discussion and document your answers to the Study Questions before reviewing the Answer section. You can reconcile differences between your own answers and the chapter Answers during additional chapter review.

Finally, this book would not have been possible without drawing on the resources and works completed by other people in their respective fields. Where references are made they will be noted in the chapter and summarized at the end of the chapter where they are used. The references will be in the author-date citation style.

Your choice is now where to begin. Review the Contents and decide. Your journey to greater knowledge in the area of solids level measurement monitoring awaits you.

Chapter 1

General Use and Applications

Objectives

After reviewing this chapter the reader should be able to:

- Define "level measurement."
- Explain the three categories of level measurement.
- Understand the differences between level measurement of liquid and bulk solid materials.

Summary

Definitions of level measurement have generally focused on the measurement of liquid material in vessels. Differences between liquids and bulk solid materials are substantial, with regard to their behavior within a vessel due to the nature and characteristics of both liquid and solid states. This means that we must identify, define, understand, and consider the unique characteristics of bulk solid materials as they exist within their containment vessels before deciding upon how best to accomplish the measurement or monitoring of their level. Uneven material surface, material flow characteristics and problems, and the seeming compressibility of bulk solids within their vessel must be properly understood and considered.

1.1   What Is Level Measurement?

The term "level measurement" can be defined in several ways. Let's look at the words themselves. The word "level" can be used as a noun, verb, or adjective. In our application it is generally used as a noun. *Merriam-Webster*'s online dictionary has several definitions; those most applicable are "a measurement of the difference of altitude of two points by means of a level" and "an approximately horizontal line or surface taken as an index of altitude." The use of the word "altitude" in the definition implies the height of something. We also see reference to a "horizontal line" and the difference in the altitude between two points. In our context "level" is the height of something as determined by the distance between two defined points parallel to each other. An example is shown in Figure 1-1 with a tank of water.

The bottom of the tank represents the lower of the two horizontal lines and the current location of the surface of the water represents the second horizontal line, which is at some altitude higher than the bottom line. The second word in our term is "measurement." This word is also a noun. It is the result of the act of "measuring," which is the action expression of the verb "measure." When we put the two words together to form the phrase "level measurement," then apply our context, we can combine the individual definitions to come up with a good definition of "level measurement": *The measurement of the value of the height of some material from a lower point to a higher point when we draw horizontal lines plumb and parallel to each other at the two points.*

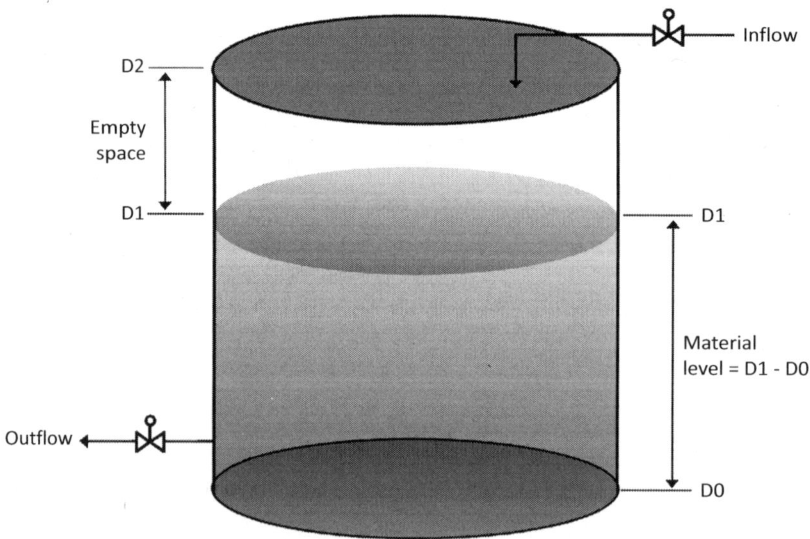

Figure 1-1   Example tank with a material at a variable level.

1.2   The Categories of Level Measurement

Our definition of "level measurement" gives rise to many possible applications in a wide range of industrial and commercial uses. It also allows us to draw some distinctions between measuring the level of a liquid and that of a bulk solid, whether powder or granular. Here we examine typical usages of level measurement and then discuss the type of material (liquid or bulk solid). There are three categories of level measurement: level indication, inventory monitoring, and process level.

Level Indication

The use of a level indicator to advise a control system, operator, or other personnel of the presence or absence of material at some predetermined point is referred to as "level indication." The most common use of a level indicator is for high, low, and intermediate level indication. These applications are also commonly referred to as "point level" or "level control." However, as we will see in our discussion of process level, level control is not exclusive to point level or level indicators.

Level indication applications are control type applications where the sensing is used to control some function or process such as high level indication and control of a filling process for a liquid or bulk solid. High level indicators detect material *presence* at a predetermined high point within a vessel. On the other hand, low level indicators detect and report the *absence* of material at some predetermined point and are used to shut down a production process or initiate a reorder or replenishment of material. Intermediate level indication is the detection of material presence or absence at some predetermined point between the high and low points. High and low indication points are usually considered emergency control points for the filling and discharge processes. Intermediate level indication is considered a demand point, calling for more material, indicating adequate level for processing, etc. Figure 1-2 illustrates these three level indication applications.

Inventory Monitoring

Level measurement can also refer to measuring and monitoring material inventory. This is common in liquid applications within the petroleum and refining industries. We also commonly see these applications in virtually every industry that processes bulk solids. Plastic processors maintain inventories of plastic resin and pellets as raw material inventory. Chemical

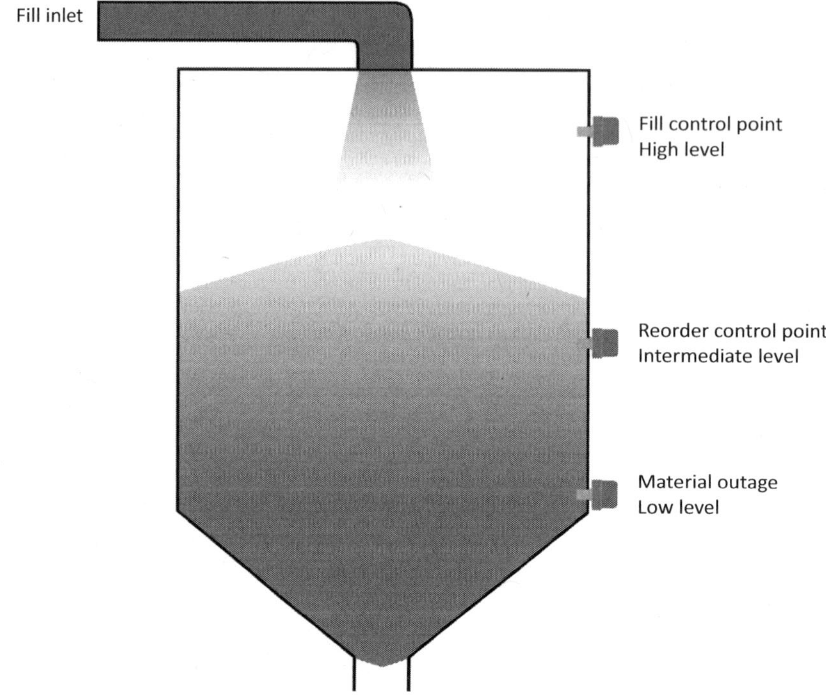

Figure 1-2   High, low, and intermediate level points.

facilities produce plastic resin and pellets for use by these same plastic processors. Food manufacturers store raw ingredients such as sugar and flour and require inventory counts and monitoring. Feed mills maintain inventories of grain and other ingredients as raw material inventory and also finished product in granular or pellet form. All of these materials are often kept in bins and silos for temporary storage. Inventory monitoring answers the question, "How much material do I have?" It is the measurement of the material level on a continuous basis with some degree of frequency or response time. These applications are not for the purpose of controlling some process or function. They perform an operational function to quantify the inventory of material for financial purposes. Inventory feeds the production process. Without an adequate amount of inventory, the required production may not take place. Inventory is also a financial asset. The value of the assets of a business impacts the balance sheet of the company and its net worth. Knowing how much inventory exists is very important. The accuracy of the inventory measurement often becomes a point of discussion or debate, as we will review in a later chapter.

Inventory monitoring applications use "continuous level" technologies and instruments to measure the inventory amount. The height of the material pile is measured on a continuous basis and updated with some degree of frequency, which is usually based on the response time of the technology used to make the measurement. Changes in material inventory as the vessel is filled or the material is withdrawn and used within the production process will cause the inventory level value to change.

Process Level

When continuous level measurement is used for control purposes, rather than inventory measurement, this is known as a "process level" application. These applications appear primarily in liquid processes. Consider the goal of controlling the flow rate of a fluid from one pressure vessel into another where the flow rate is required to remain constant (Wallace, 1980). Because the two vessels are pressurized and liquids are generally incompressible, the flow rate will fluctuate when the pressure in either vessel changes. In order to maintain constant flow the level measurement in the feed vessel is used as a set-point input to the control loop working to maintain the feed flow to the receiving vessel. This level measurement in the feed vessel is a continuous level measurement, typically providing a variable output such as a 4–20 mA signal. Process level measurement applications of this type are not commonly used in bulk solid processing as most vessels are vented (not pressurized). However, process level applications are not completely exclusive to liquid applications. One example of a process level application with bulk solids is the use of a surge bin within a process where the surge bin level is measured and used to help control or bring balance to a production process.

1.3   Differences Between Liquid and Bulk Solid Levels

We now have our definition of level measurement and an idea of how measuring and monitoring the level of material in vessels can be used by a wide variety of industries. Now let us look at the key characteristics of bulk solid applications and how they differ from measuring liquid levels.

We defined the term level measurement as *the measurement of the value of the height of some material from a lower point to a higher point when we draw horizontal lines plumb and parallel to each other at the two points.* This works well for liquids because of four key properties associated with them: (1) liquids flow readily and are known to "seek their own level," (2) liquid particles move readily, (3) the surface of a liquid in a vessel will always be "level" or flat, and (4) liquids are generally considered to be incompressible and do not "pack." The last statement is not technically true, but based on the amount of pressure required to compress a liquid, and considering the very small amount of compression that would result, liquids are virtually incompressible. Considering these four points, measuring the distance from the top of a vessel to the surface of the liquid is the same at one place on the surface as it is at any other place on the surface of the liquid. In fact, doesn't the very definition of level measurement imply that it is strictly meant for liquid material?

Now let's take a look at bulk solid materials. Measuring the level of a bulk solid material in a vessel presents challenges due to the unique characteristics and behavior of the material in solid form. These include: (1) most bulk solid materials have a particle size visible to the eye or at least significantly larger than a liquid particle, (2) the air between particles allows bulk solids in vessels to behave in a compressible manner, and (3) significant friction exists between solid particles, as well as between the solid particles and the walls of the vessel. This means that bulk solids will not flow like a liquid and the surface of the bulk solid is more likely to be irregular rather than flat. It also means that the volume of material can change without adding or taking away any material. The latter is especially true of powders that are pneumatically fed into their storage vessels. The air entrained with the particles during filling escapes as the material settles, even if no material is added or discharged. But why do these characteristics and behaviors exist? This is related to "movement."

Liquid particles move at a rapid rate, compared with solids, but slower than gas particles (Whitcomb, 2010). Solid particles at the molecular level basically do not move; therefore, a solid particle will not assume the shape of its container whereas a liquid particle will. In addition, solid particles have a strong attraction to each other. Liquid particles tend to move past one another easily. Further discussion of bulk solid flow properties can be found in Appendix B. It is enough to say at this point that the difference between liquid and solid particle movement and attraction has a dramatic impact on level measurement applications. Table 1-1 summarizes these differences.

The flowability (Prescott and Barnum, 2000) of a bulk solid material presents the features of a level measurement application that need to be considered when looking for a level measurement device. Let's take a closer look and summarize these features: (1) the surface of the material will be irregular, not flat like a liquid material; (2) the pile of material can settle during and

Table 1-1   The Difference Between a Liquid and Solid Particle

	Liquids	Bulk Solids
Compressibility	Liquids require extremely high pressure to compress and the amount of compression is small. Liquids are incompressible from a practical standpoint. They will take and hold the shape of their container.	Solid particles at the molecule level are incompressible. *Note*: Bulk solids in vessels may appear "compressible" because the air entrained within the bulk solid mass will escape as the bulk solid material "settles" within the vessel.
Particle speed	Virtually undetectable to the naked eye because liquid molecules move fast compared with solids, though much slower than a gas. The visible particle size therefore appears to be much smaller than solids.	Bulk solids have molecules that do not move. They do not assume the shape of their container like liquids do. They do not flow as well as a liquid.
Attraction	Liquids move rapidly and are not attracted to each other in comparison with solid particles.	Solid particles are attracted to each other and to their container. This means, in comparison with their liquid counterparts, solid particles do not flow or will not flow as well as liquids.

after filling due to the elimination of air as the pile increases in mass and the material settles; (3) the bulk solid material will not assume the shape of the vessel; however, the material pile will shift and move due to gravity as the weight of the material overcomes the friction holding solid particles together; and (4) because solid particles tend to cling to one another, this "friction" between solid particles and the vessel walls can result in sidewall buildup or buildup of material on any invasive sensing element, even in the case of fine powders or dust.

The shape of the material surface of a bulk solid in a static condition is known as its "angle of repose." This "angle" is not necessarily a single or even constant angle across the entire material surface. As the bulk material fills a vessel, the angle of repose typically creates a peak (see Figure 1-3).

Conversely, when the material is discharging it will be drawn down and transition into the shape of a valley. There is a third condition, this being the transition point when the material surface moves from a peak to a valley. The most consistent and repeatable angle of repose will typically exist when the vessel inlet is located directly over the discharge, and both are in the center of the vessel. This is important to understand when using level measurement for inventory monitoring purposes because traditional continuous level measuring devices measure the distance to a single point on the surface and conversion algorithms used to calculate volume or mass assume a flat surface. Choosing the mounting location for a high, low, or intermediate level sensor is also impacted by the angle of repose. As an example, consider a level indicator used for control of the filling process. The high point needs to consider the run-out from the filling system and the mounting point where the material will be detected on the angle of repose.

Bulk materials move and shift during filling and discharge due to their seeming compressibility. The forces imparted upon invasive sensors by the

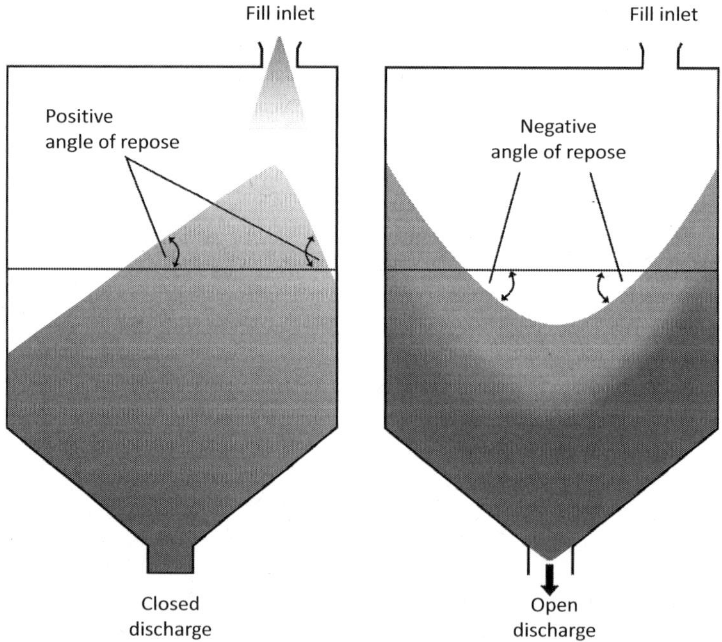

Figure 1-3   Angle of repose.

weight of the bulk solid material as it flows and moves can be significant. For example, manufacturer installation, operation, and maintenance instructions (Monitor Technologies, 2011) for a guided wave radar continuous level sensor indicates that the "traction force" or tensile loading imparted upon a 8 mm diameter stainless steel sensor cable 65 feet (19.8 m) long can be as much as 2 metric tons. In fact, these sensor cables are typically designed to withstand over 3 metric tons in loading.

Some powder materials stick together or to the vessel walls due to their nature or because of their particle attributes and coefficient of friction. Sidewall buildup and rathole: these occur as a result of problems related to bulk solid material flow and friction between material particles and the vessel. These material flow issues can impact proper operation and performance of both bin level indicators and inventory level measurement sensors. Bin level indicators mounted on the side of a vessel may become covered by material buildup on the wall of the vessel, while the material in the center, or other areas in the vessel, has completely emptied. This could present a false indication of material presence or add error to the inventory measurement.

1.4   Conclusion

Measuring the level of bulk solid material presents application characteristics very different from measuring the level of liquids. These differences demand our attention if we are to select the best level measurement sensor device to meet the objectives of each specific application. These differences are related to the unique properties of solids regarding their movement and behavior within a vessel. Because solids do not move well and because they adhere to each other, the flowability of bulk solids is problematic in comparison with that of a liquid. Next, we will look at how these, and other, application characteristics impact the selection and successful use of technologies for point level detection and indication.

Study Questions

Q1:   What are the two primary types of level measurement devices, their primary function, and application category or type?

Q2:   Explain the fundamental difference between liquid and solid particles and what this means to us for level measurement applications?

Answers

A1:   Level measurement devices are classified as either point level or continuous level. Point level devices are designed and used to detect the presence and absence of material at some predetermined point chosen by the user. Continuous level devices detect the continuously changing level of material with some frequency based on the device technology.

Point level devices are used for control functions, such as control of a vessel filling operation, to control replenishment of material, or to control some aspect of a production process downstream or upstream from the point of measurement. Point level devices fall under the category of level indicators as they are almost strictly used for control. The only exception is the use of point level indicators for rough estimation of material inventories when mounted spaced apart by some distance on the vessel. This is explained further in the next chapter on point level detection.

Continuous level devices for use with bulk solids are used primarily for inventory monitoring functions. In this application the continuous level measurement device is used to determine and report "how much" material is within the vessel on a periodic basis. In these cases the continuous level device falls into the category of being an inventory monitor. However, there are applications where the continuous level device may be used in controlling a process such as surge control. In liquid level applications these devices can be used to control the flow of fluid between pressure vessels.

A2:   Liquid particles move, solid particles do not. This is what allows a liquid to flow and take the shape of its container, while solids do not readily do this. This creates irregular surfaces in solid materials within vessels, as well as buildup and other flow problems during discharge. Liquids, for the purposes of our discussion, are not compressible and because they move, a liquid will nicely fill its vessel. Solid particles are not compressible either; however, in a vessel they can give the appearance of compressing as the pile of a bulk solid material settles and air escapes from between particles. The more air escapes, the more the pile of solid behaves as a single mass, which is difficult to flow. In attempts to solve bulk solid material flow problems people can sometimes resort to violent behavior as they attack the vessel with hammers, manual or automatic types (McGregor, 2010).

References

McGregor, Robert G. 2010. A cure for hammer rash: Measuring powder flowability with shear cell testing. *Powder and Bulk Engineering* 24 (September): 37–44.

Monitor Technologies. 2011. *Flexar® guided wave radar continuous level measurement system*, p. 7. Bulletin 354A, DOC-001-354A Rev. 5, dated June 17, 2011, Elburn, IL: Monitor Technologies LLC.

Prescott, James K., and Roger A. Barnum. 2000. On powder flowability. *Pharmaceutical Technology* (October): 60–84.

Wallace, Leonard M. 1980. Sighting in on level instruments. In *Practical process instrumentation and control* (Chemical Engineering), 253–62. New York, NY: McGraw-Hill Publications.

Whitcomb, Todd W. 2010. Solid state. In *Chemistry explained: Foundations and applications*. www.chemistryexplained.com/Ru-Sp/Solid-State.html (accessed October 4, 2010).

Chapter 2

Point Level: An Introduction

Objectives

After reviewing this chapter the reader should be able to:

- Define "point level" and describe its primary function.
- Discuss the process control functions served by point level measurement devices.
- Explain the principle of operation for six point level measurement technologies.

Summary

We defined level measurement in Chapter 1 and identified two types of level measurement devices, that is, point level and continuous level. Point level devices are used primarily for control purposes and are therefore sometimes referred to as level controls. Controlling a function or aspect of a process based on point level detection in modern times dates back to the early part of the twentieth century when process measurement and control instrumentation was in its infancy. Mechanical and electromechanical devices were the first to emerge and many of these early technologies continue to be manufactured and used today. However, several electronic or solid-state devices have been introduced within the past few decades that have supplanted some of the mechanical and electromechanical units. Irrespective of the technology, the devices are used to detect material at predetermined points for the purpose of controlling a function or process with their output. Knowing how to differentiate between the technologies will allow for a proper understanding of selecting the best technology to fit a specific application.

2.1   What Is Point Level Measurement?

We defined the term "level measurement" in Chapter 1 as the measurement of the value of the height of some material from a lower point to a higher point when we draw horizontal lines plumb and parallel to each other at the two points. We also indicated that the term "point level" is synonymous with the terms "level indication" and "level control." Therefore, point level measurement is an application where detection of presence and/or absence of a material at a predetermined point is done for the purpose of controlling some function or process.

Determining the point at which the detection is needed, and whether the critical measurement is the presence or absence of the material, depends strictly on aspects of the control function. Most level control functions can be grouped into two categories, high and low. It is quite simple: Detecting material when it exceeds a certain point is high level and when it falls below a point within a vessel is low level (see Figure 2-1).

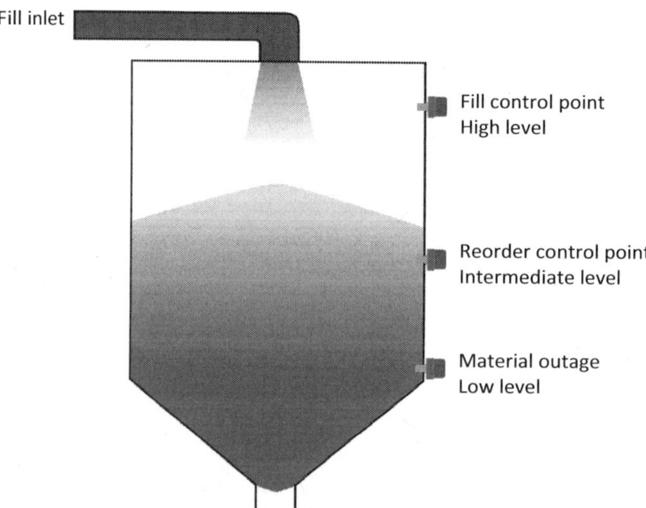

Figure 2-1   Typical control function categories.

Table 2-1   Typical Point Level Application Objectives

High Level	Low Level
Prevent overfilling	Trigger replenishment of material
Fill vessel to capacity	Prevent process starvation and shutdown

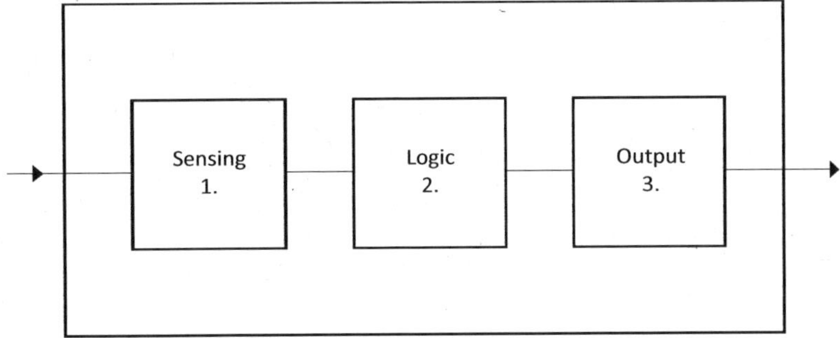

Figure 2-2   Primary function blocks within any point level sensor.

In a high level control or point level application the critical detection is the "presence" of the material at the predetermined high level point. In a low level control or point level application the critical detection is the "absence" of the material. These are very common applications. The primary objectives of each are shown in Table 2-1.

A point level measurement sensor has three function blocks as shown in Figure 2-2. No matter what technology is utilized, each point level sensor has a way to (1) sense the material, (2) use some form of mechanical, electromechanical, or electronic logic to differentiate between presence and absence, and (3) indicate through an electrical output the state of the device alarm or material condition (present or absent).

2.2   A Word About "Process Control"

The function of process control in our discussion is "feedback control." Feedback control can be automatic or manual. In either case, feedback control always compares a measured variable with a setpoint and then manipulates some final element within the process to effect a change in the actual condition until it meets the desired setpoint (see Figure 2-3).

In our example we wish to control the shutoff of the flow of material into a storage silo. Our objectives, shown in Table 2-1, are to prevent overfilling and to fill the silo to its maximum capacity. The control setpoint is the actual

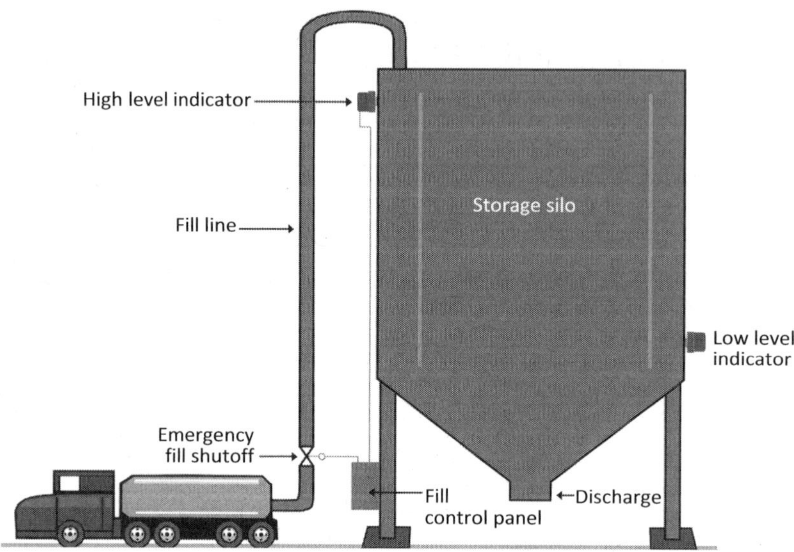

Figure 2-3   Automatic feedback control when filling a vessel.

level of the material in the silo considered to be the shutoff point so that when the material remaining within the fill transport finishes emptying into the silo it will be filled to its maximum capacity. The setpoint is the location of the sensing element portion of our point level sensing device within the silo. In this example the point level sensor is mounted on the side of the vessel; however, high level sensors can also be mounted on top of the vessel. The output from the high level sensor provides an electrical contact closure connected with the automatic filling control system. The electrical contact closure is the control system feedback indicating that the material in the silo has reached the setpoint. The filling control system then shuts off the conveying system. This is an automatic feedback control system. A manual control might use the point level sensor to detect the presence of the material and indicate the high level condition by powering an alarm light and horn, thereby requiring intervention of an operator to shutdown the filling system. In either case, automatic or manual control, the point level sensor is a key component in the process control system. It represents the measurement of the variable and its output is the input to the control system. Its performance, accuracy, precision, response time, and reliability are critical to the control function.

Point level sensors are provided with either an electric switch or a relay contact as their output device. Terminology to describe the condition of the electrical contacts in the switch or relay includes normally open and normally closed. With an electrical switch these terms describe the condition of

Table 2-2   Contact Arrangements in Level Controls

Contact Arrangement	**Description**
SPST (single pole, single throw)	Form A—one normally open contact Form B—one normally closed contact
SPDT (single pole, double throw)	Form C—one normally open contact AND one normally closed contact
DPDT (double pole, double throw)	Two Form C—two normally open contacts AND two normally closed

the contacts during normal (not alarm) conditions, unless specifically mentioned otherwise by the manufacturer. Contacts that are normally open will go closed, allowing electricity to flow across the contact, when the alarm condition occurs. Conversely, normally closed contacts will go open, breaking or stopping the flow of any electricity that might have been flowing across the contacts. With a relay, normally open or normally closed describes the condition of the contacts when the relay is de-energized (no power is flowing to the relay coil). The switch or relay contacts allow a remote device to be turned on or off. The remote device could be an alarm light, siren, motor starter, or the contacts can be used to open or close a valve. Process equipment or devices are often directly activated due to the high current rating typical of these contacts (as low as 2 A to as high as 20 A). The point level sensor contacts are often arranged together. For example, an electrical switch or relay may provide multiple contacts being activated simultaneously. Refer to Table 2-2 for a list of the most common contact arrangements used within point level sensors as their control output.

Remember the old computer saying garbage in, garbage out? Selecting the most appropriate point level sensor for the application can make all the difference. We will talk about the selection process in Chapter 3, but next we will identify and discuss the common technologies that can embody point level sensors for use with powder and granular bulk solids.

2.3   Point Level Sensor Technology

The use of modern day technology in point level measurement for bulk solids arguably dates back to the early part of the twentieth century when on February 27, 1918, a patent application was filed by Edson C. Covert of New Kensington, PA, titled "Indicator for Storage-Hoppers or Other Receptacles." The patent was granted and issued as US Patent 1,379,803 on May 31, 1921 and assigned to Heyl & Patterson, Inc., of Pittsburgh, PA. According to *Wikipedia* (2010) Heyl & Patterson was founded in 1887 and by the time this patent was filed and issued they were a leading engineering and manufacturing company for bulk solid material handling, transport, and storage. This patent ushered in the diaphragm pressure-sensitive level switch. Note the diaphragm material listed within the patent text is stated as "asbestos cloth provides a suitable material and the diaphragm illustrated is made up of such asbestos cloth." Not a suitable material today. In addition to the pressure-sensitive diaphragm technology we will discuss tilt, rotary paddle, RF admittance/capacitance, proximity switch, and vibrating element technologies to understand their background and generic principle of operation.

Pressure-Sensitive Diaphragm

This material detection technology utilizes a mechanical system which activates and deactivates an electric switch. No power is required for this type of point level switch to function. The mechanical system (sensing and logic blocks as shown in Figure 2-2) moves in response to material density against the diaphragm (see Figure 2-4). The diaphragm isolates the internal components from the material inside the vessel. When an adequate amount of material is present inside the vessel and accumulating against the diaphragm, the mechanical system (sensing and logic block functions) is put in motion, which in turn activates the electric switch (output block function) and changes the condition of the electric switch contact(s). Contacts that are normally open go closed and those that are normally closed will go open. The mechanical system can have a fixed or adjustable sensitivity (part of the logic block function) that will determine the minimum bulk density of material that can be sensed and can activate the switch's mechanical system. Often the sensitivity, if adjustable, is set by an adjustment screw, which increases or decreases the mechanical system resistance to the force applied upon the diaphragm by the bulk material inside the vessel.

The primary functional advantage of the pressure-sensitive diaphragm level switch is in its very limited invasiveness into the vessel. There are limitations and disadvantages as well, primarily that this level switch is limited to use with very dry, free-flowing material, and has moving parts prone to failure due to wear and tear. A more comprehensive list of the pros and cons of the pressure-sensitive diaphragm level switch is shown in Table 2-3.

Figure 2-4 A commercial form of the diaphragm level switch (Diaphragm switch brochure 02-11-1M-NPC, page 2, Model BM45, BinMaster, Lincoln, Nebraska).

Table 2-3 The Pros and Cons of the Pressure-Sensitive Diaphragm Level Switch

Pros	Cons
Minimally invasive (very low profile). Can be considered where invasive sensors may be problematic.	Cannot use with material that may stick or build up on the diaphragm in any manner. Mount on top or vertical straight-wall only. Any material resting on or adhering to the diaphragm can create false indication of true material presence.
Can be mounted from outside the vessel. Simple to install.	Moving parts means the mechanical system is subject to wear and tear based on frequency of activation and operating environment.

(*Continued on following page*)

Table 2-3 (Continued)

Current rating of switch is very high (15 A is typical) and can turn process equipment on/off directly.	Diaphragm is sensitive to damage due to abrasion or sharp and heavy material with moderate to large particle size.
Low purchase cost.	Typically limited to moderate operating temperatures: <+200°F (+94°C); optional diaphragm materials up to 250°F (120°C).
Very simple mechanical level switch means it is easy to operate, doesn't require power at the sensor mounting location, and is easy to maintain.	
Moderate life expectancy when applied and installed properly.	

Tilt

A tilt type device is a level switch based on detecting the presence of the material as it becomes tilted by a pile of material as it increases in height and contacts the level switch enclosure. Absence of material is detected as the tilt switch returns to a plumb condition. The activation tilt angle can vary depending upon the manufacturer. However the tilt angle is typical 15–17° from plumb. The tilt switch is used only for high level indication and is installed hanging or suspended over the material at a point where it is not in the fill stream but will come in contact with the material as the pile increases in height. The tilt level switch can be used within enclosed vessels, in open pits, or with open-top vessels (see Figure 2-5).

Early tilt type level switches were based on mechanical activation (sensing and logic block functions) of an electric switch (output block function) at a predetermined tilt angle. In some brands the mechanical activation system was replaced with a mercury-switch (logic block function only). Mercury switches (logic block) are a glass assembly with two poles or electrodes within one end of the assembly. Liquid mercury is within the glass assembly but does not completely fill it. As long as the position of the tilt switch is such that the liquid mercury remains at the opposite end of the mercury switch assembly from the two electrodes or poles, no connection between the poles is made. As soon as the tilt switch

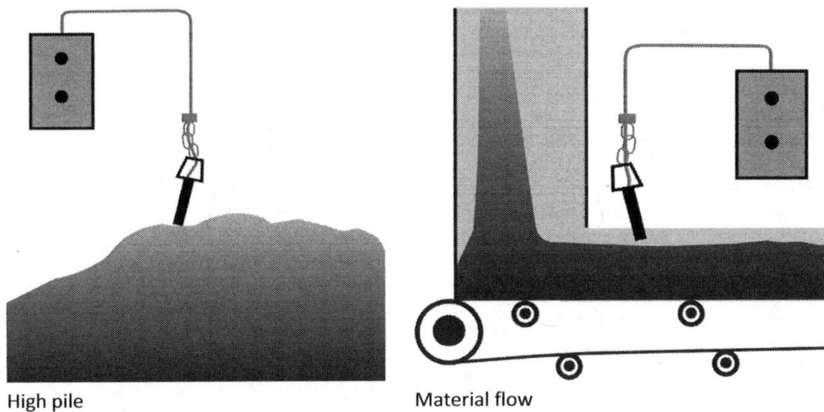

Figure 2-5   Typical applications of the tilt level switch include high levels of material piles and over/underloading on belt conveyors.

reaches its activation angle the mercury moves to the other end of the mercury switch tube and makes an electrical connection between the two poles or electrodes. When the tilt switch returns toward its original plumb position the liquid mercury moves back to the end of the mercury switch assembly and the circuit between the two electrodes is broken. Mercury switches are commonly used in older thermostats for central heating and cooling systems. They are tilt activated electrical switches. A disadvantage of the mercury switch based units is that mercury is a liquid metal and is considered an extremely hazardous material. In tilt switches using mercury filled switches they are completely encapsulated within a heavy-duty probe, typically made of heavy steel. However, some may question the method of disposal of these switches as they still contain this toxic material. In recent years tilt level switches have become available that replace the mercury switch with an optical or other solid-state tilt sensor (sensing block function) and associated electronics (logic block function) such as the unit shown in Figure 2-6.

This provides an alternative to both mechanically and mercury switch activated units. The primary advantage of the tilt level switch is its low purchase cost, when used without a control unit, in addition to these switches not requiring a power supply for the mechanically activated units. However, these level switches are limited to high level applications and are not suitable for use with lightweight materials; typically the material bulk density must be ≥ 15 lb/ft^3 (240 kg/m^3). A more comprehensive list of the pros and cons of tilt switch technology is shown in Table 2-4.

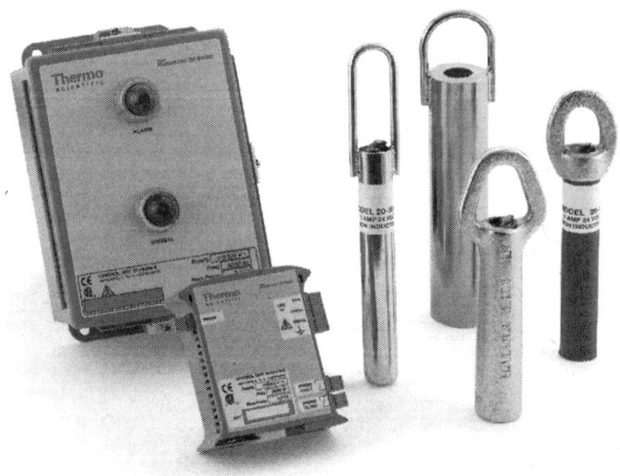

Figure 2-6   Non-mercury tilt switch and controls (Mercury-free tilt sensors, product specifications, Thermo Fisher Scientific, Inc., Minneapolis, Minnesota).

Table 2-4   The Pros and Cons of Tilt Switch Technology for Point Level Sensors

Pros	Cons
Low purchase cost for mechanically activated tilt switches or for probe only mercury switch based installations.	Limited to high level applications only. Must be suspended over material at desired high level activation point.
No power required for mechanically activated units. The only electricity in this type of tilt switch is that wired through the electrical switch inside the unit.	Units based on mercury switch activation use hazardous toxic material.
Current rating of switch is high (10 A typical) and can turn process equipment on/off directly	Tilt probe unit may be subjected to wear and tear from material, especially heavy aggregate material with large particle sizes.
Solid-state tilt switch units eliminate mechanical moving parts and mercury switches.	Typically limited to moderate operating temperatures; < +200°F (+94°C).
Moderate life expectancy when applied and installed properly.	

Rotary Paddle

Originating at least as far back as the mid-1930s to a patent by Herbert S. Lenhart (1936) assigned to Fuller Company that was applied for on November 28, 1936, and issued on May 3, 1938, the rotary paddle point level control device operates on the principle of material resistance to a rotating blade, which has a defined surface area, commonly known as a "paddle." The paddle (sensing block component) rotates freely in open air within the bin driven by a small motor within the level sensor enclosure. The paddle is attached to the rotating output shaft protruding from the process connection of the level control enclosure (see Figure 2-7). As the material comes into contact with the rotating paddle, the rotation is impeded. The rotary paddle point level sensor is an electromechanical device. The internal electromechanical system (logic block function) controls the rotation of the paddle and detects material presence and absence.

Typically the drive motor is attached to a mounting plate that is free to rotate within the enclosure. When the material impedes the paddle rotation, the continued operation of the drive motor causes the motor and mounting plate assembly to rotate on its axis (the output shaft linked to the motor). This rotation will trip electric switches (output block function) as the direct output of the device.

In some designs the motor stalls at the end of the motor mounting plate assembly rotation. Other designs will use the rotating motor mounting plate to trip another switch that is used to cut off the power supply to the drive motor. A note about the drive motors: The drive motors used in rotary paddle bin level indicators are typically synchronous motors. There are two types, the hysteresis and permanent magnet types. Those rotary paddle units that use the hysteresis type will stall the motor when the material impedes

Figure 2-7 Typical rotary paddle point level sensor with paddle attached (Courtesy of BlueLevel Technologies, Inc., Rock Falls, Illinois).

the paddle rotation. This is acceptable with hysteresis motors so long as the gearbox design internal to the motor assembly can withstand the stalled condition. Permanent magnet types of synchronous motors are not designed to be stalled and therefore must be shut off soon after being in a stalled condition.

The rotating mounting plate assembly is spring-loaded. Therefore the assembly will rotate in the reverse direction back to its origin when material recedes from the paddle. This will reengage power to the drive motor (if equipped with permanent magnet motors and a motor shutoff) and reverse the state of the electric switch output.

While some manufacturers use electric switches for the device output it is also common to use a microswitch to activate the coil of an electromechanical relay and the contacts in the relay become the output from the rotary paddle level sensor. This allows the design to de-energize the relay coil when in the alarm condition so that any power failure to the device will also create an alarm condition in the relay contacts, thereby triggering an alarm upon power failure. The output therefore is deemed "fail-safe" on power failure conditions, meaning that the output of the level sensor is deemed to be in the safest position (alarm) if power to the sensor fails.

Rotary paddle technology has also been used to produce a level sensor control device that has the ability to detect the presence of electrical or even mechanical failures (see Figure 2-8). This unit is typically more expensive

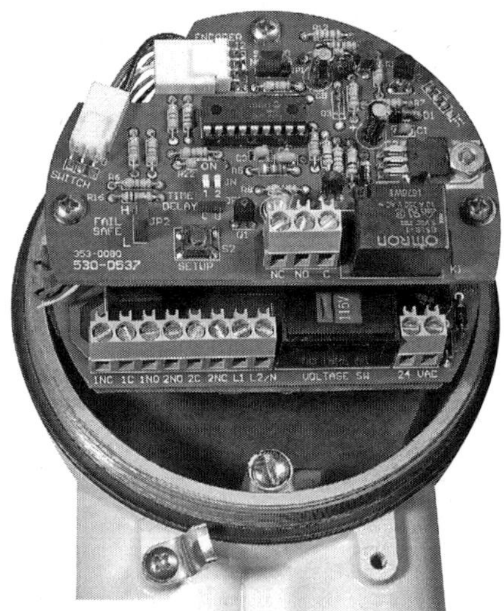

Figure 2-8   Internal view of self-validating rotary paddle point level sensor (MAXIMA+ Rotary Level Indicators brochure, page 4, BinMaster, Lincoln, Nebraska).

than the standard rotary paddle devices previously discussed, but provides indication of a device problem when it is detected. It is not "predictive," but it does "self-validate" its ability to operate properly on a continuous basis and informs an operator or control system of that status, in addition to being able to detect material presence or absence at the predetermined point chosen within the vessel.

The rotary paddle point level sensor has the distinct advantage of being near-universal in its ability to work with powder and bulk solid materials. The paddle attached to the output shaft of the unit can be changed to adapt the sensor to very heavy or very lightweight material. Most material characteristics do not affect the units' operation. However, the rotary paddle level control unit is an electromechanical device with moving parts subject to wear and tear over time. The electric drive motor is usually the most common point of failure for these units, but is easy to replace and generally low in cost. The pros and cons of rotary paddle technology are summarized in Table 2-5.

Table 2-5 Pros and Cons of Rotary Paddle Point Level Sensor Technology

Pros	Cons
Low purchase cost compared with most solid-state point level sensors.	Electromechanical nature means moving parts subject to wear and tear over time.
Near universal in ability to detect bulk solid materials, from 2 to 200 lb/ft³ (32–3200 kg/m³); fine powder to heavy material and even materials with large particle size; use as low, high, and intermediate level control.	Drive motor is most common failure mode due to its sensitivity to start/stop, heat, and wear and tear during continuous operation.
Easy to install and simple to maintain.	More expensive than other mechanical level switches for some applications.
Wide range of options and accessories.	
Proven with over seven decades of refinement.	
Self-validating version available that detects and indicates failure conditions.	

Capacitive Technology in Proximity Sensor Form

Capacitive proximity technology has been used as a point level sensor and target proximity sensor for many years. US patents date back at least to the mid- to late 1960s. Today a capacitive proximity switch is a completely solid-state device. The absence of moving parts makes it attractive, as well as its small size and low purchase cost. Let us examine the basics of how these devices work (see Figure 2-9).

Capacitive proximity sensors generate an electrostatic field and detect changes that are caused by introduction of a target or the material within the established field. As the material comes within the field the capacitance of the electronics will increase. The oscillator of the capacitive proximity sensor becomes active as the capacitance exceeds the threshold, which triggers the sensor output. The threshold at which the output is activated is based on an adjustable setting within the sensor. This adjustment is typically known as the distance adjustment or sensitivity setting. The capacitance generated by the sensor electronics is based on three key factors. These are: (1) the proximity or closeness of the target to the active field, (2) the size of the target, and (3) the dielectric constant of the target. The closer the target is to the sensor, the greater the size of the target coming within the field generated by the sensor; or, the higher the dielectric constant of the target/material, the larger the capacitance level generated within the sensor electronics will be. An adjustment potentiometer is typically located on the back of the capacitive proximity switch allowing for regulation of the operating distance, sensitivity, or threshold. This is useful in most applications, such as with the detection of full or empty conditions in hoppers, tanks, and bins. The sensitivity can be adjusted so that the material is sensed at whatever distance is required or upon contact of the material with the front face of the sensor. As the generated field

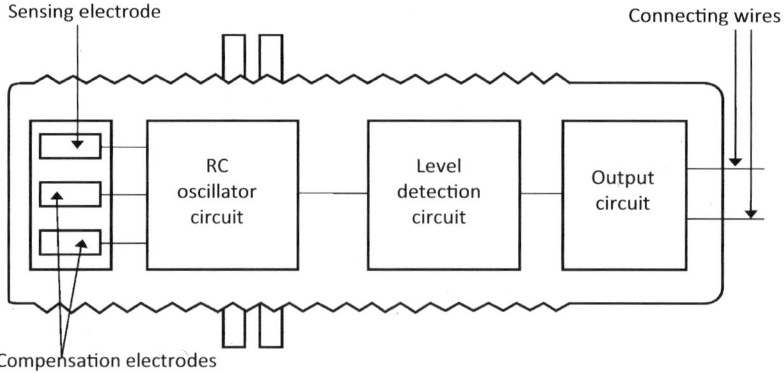

Figure 2-9   Block diagram of capacitive proximity sensor.

is fixed, capacitive proximity switches will have a maximum sensing distance that can be adjusted from some minimum to a maximum (typically 12–25 mm), which will vary from brand to brand.

Two types of capacitive proximity switches exist: shielded and unshielded. The following paragraphs briefly explain each and their typical use.

Shielded Sensor Construction: "Shielded" sensors are made with a metal band or shield surrounding the sensor probe (see Figure 2-10). This focuses the electrical field with a straight-line path in front of the sensor resulting in a more concentrated field. The primary advantage of shielded sensors is that they can be mounted flush in surrounding material (vessel wall) without detecting the vessel wall (Figure 2-11). Unshielded sensors cannot be flush mounted, as they would detect the vessel wall, thereby falsely indicating detection of the target material.

Figure 2-10   Diagram of shielded capacitive proximity sensor.

Figure 2-11   Shielded capacitive proximity sensor (Courtesy of BlueLevel Technologies, Inc., Rock Falls, Illinois).

Unshielded Sensor Construction: Unshielded sensor construction does not have a metal band surrounding the probe and these sensors have a less concentrated field than shielded sensors (see Figure 2-12).

Unshielded sensors therefore have a spherical electrical field. Because of the shape of the field these sensors are designed to touch the target product, such as a liquid or bulk solid material, including plastic pellets, sugar, flour, corn, sand, oil, and water. Because unshielded sensors are not suitable for embedding the circumference of the front face within a metal wall, bracket etc., the threading on the sensors does not extend to the front face of the sensor as shown in Figure 2-13.

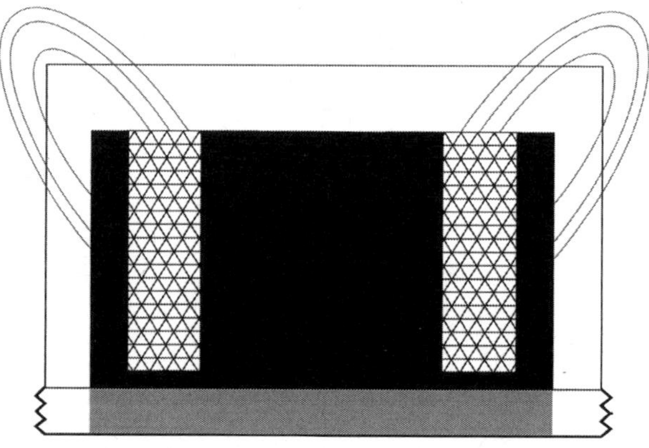

Figure 2-12   Diagram of unshielded capacitive proximity sensor.

Figure 2-13   Unshielded capacitive proximity switch (Courtesy of BlueLevel Technologies, Inc., Rock Falls, Illinois).

Capacitive proximity switches are also available based on the voltage of the load to be switched, these being AC or DC voltages. AC capacitive proximity switches will operate with some universal range of voltage, such as 20–250 V AC, and are provided with 2-wire configuration. They can switch current loads typically of only a maximum of 500 mA, and this will vary by brand.

DC capacitive switches are a little different than AC switches in that they can use one of two types of transistors as the switching output. These two types of transistors are NPN and PNP. NPN is a transistor output that switches the common or negative voltage to the load; the load is connected between the sensor output and the positive voltage supply. PNP is a transistor output that switches the positive voltage to the load; the load being connected between the sensor output and the voltage supply common or negative. Wire configurations for DC switches can be 3-wire NPN, 3-wire PNP, 4-wire NPN, and 4-wire PNP. Switch types can be normally open (NO) or normally closed (NC). The 4-wire DC switches allow you to wire NO and NC.

The advantages of the capacitive proximity switch generally lies in its size and economy of scale in production, as proximity sensors are extraordinarily common within automation systems. Refer to Table 2-6 for a summary of the pros and cons of these devices.

Table 2-6 Pros and Cons of Typical Capacitive Proximity Switches

Pros	Cons
Very low purchase cost.	Generally not suitable for use with materials that will cling, coat, or build up on the sensor face. Use with clean liquids and dry free-flowing solids.
Very compact, the most common size is 30 mm in diameter and generally 80–100 mm long.	Low current load handling capability—typically 200–500 mA range.
Easy to install and setup.	Not as sensitive as other solid-state sensors.
"Throwaway" maintenance—highly reliable, many carry long-term warranty.	Requires calibration/tuning to application.
Low profile or flush mounting.	

RF Admittance/Capacitance

In electrical engineering terms admittance is the inverse of impedance (1/Z = Y, where Z is the impedance measured in ohms and Y is the admittance), which is the resistance to the flow of current within an AC circuit. The term RF is an acronym for radio frequency. As a type of electromagnetic radiation, radio waves are generated by electrical signals which oscillate in a specified range between approximately 9 kHz and 300 GHz. However, it is possible that a variant of this technology can be implemented where the electromagnetic radiation is at a frequency even lower, below the traditional "radio" band. Most manufacturers use technology radiating within the radio band, typically around 100 kHz.

Therefore we can conclude from the generic name of this point level sensor technology that this device generally involves the use of RF-based electromagnetic energy and the measurement of the admittance of an AC circuit as it changes in response to the presence of a material surrounding the probe element. What about RF capacitance level sensors? Are these the same as RF admittance point level sensors? RF capacitance based point level sensors detect changing capacitance rather than changing admittance values. Capacitance is the ability of some body or material item to hold an electrical charge. Either way, the principle is essentially the same. However, early versions of this technology were referred to simply as "capacitance sensors" and were prone to falsely indicating material presence when coating or buildup occurred along the probe and bin wall. Almost all RF admittance units today embody additional circuitry generically known as a "guard" or "shield" which provides significant resistance to the effects of material buildup. This circuitry prevents the RF signal return to ground at the vessel wall along the probe itself. The only return path to ground is through significant material presence around the probe (see Figure 2-14).

A typical RF point level sensor operates by applying a RF signal to the active section of the probe for the target material to contact. As contact occurs, the admittance in an electronic circuit increases. The active probe section and grounding at the vessel wall act as the two plates of a capacitor. The material to be detected is likened to the dielectric material between the capacitor plates and has a dielectric constant. The active section of the probe is separated from the grounding by an insulator. When there is no material present at the active probe section, the air surrounding the active section is the dielectric with a constant of 1.0. When material enters the bin and accumulates surrounding the active probe section the air

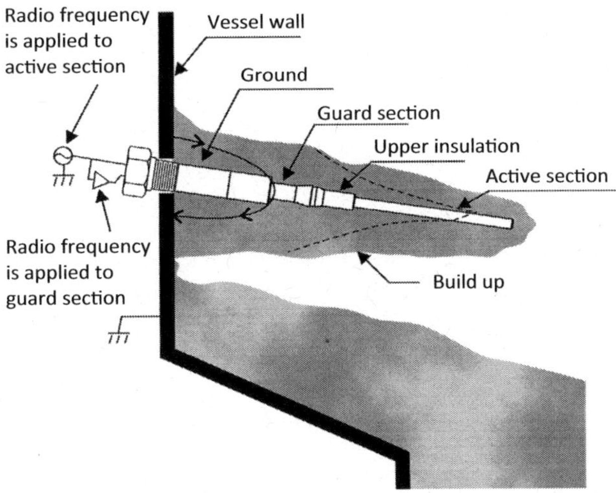

Figure 2-14   RF point level sensor probe illustration (RF-capacitance/admittance level switch data sheet, page 1, FineTek Co., Ltd., Taipei, Taiwan).

is replaced by the material with some dielectric constant greater than 1.0 and the admittance or capacitance in the electronic circuit changes indicating material presence. The RF admittance/capacitance point level sensor requires calibration, which thereby determines the magnitude of change in the electronic circuit that must exist before the material detection is determined. As shown in Figure 2-14 the probe has a guard or shield section, which also has an RF signal applied to it. This enables the device to ignore product buildup on the probe as shown by forcing the active section RF signal to return to the grounding point at the vessel wall by going around the guard influence and through air or the actual material present within the vessel.

Capacitance technology based point level sensors were developed in response to the growing need to eliminate moving parts and the maintenance associated with mechanical and electromechanical technologies even while the manufacturers of these devices evolved and improved their overall reliability, especially the rotary paddle bin level indicator. Capacitance technology was developed for liquid level applications as far back as the early 1940s. The use of this technology has grown over the years and the technology has been refined. The pros and cons of the RF admittance/capacitance devices are shown in Table 2-7.

Table 2-7   Pros and Cons of the Typical RF Admittance/Capacitance Point Level Sensors

Pros	Cons
Solid-state instrument, no moving parts.	Requires calibration/tuning to adjust the sensor to the specific installation.
Use with powder, granular, and liquid/slurry materials.	Sensitive to dielectric constant of target material, is limited to the low end of dielectric, and needs to be tuned to specific material and installation; therefore lightweight bulk solid materials and changing bin contents can be problematic requiring recalibration or changing of sensor technology.
Easy to install and set up.	More expensive than some other technologies, especially mechanical, electromechanical, and capacitive proximity switch devices.

Vibrating Element Technology

The most recent technology on the scene for monitoring the presence and absence of powder and bulk solids is the vibrating element. These devices are also known as vibrating rod, vibrating fork, or tuning fork. The last name is due to the resemblance of some units to that of an actual "tuning fork." The use of this technology was derived from its first use in determining material density in liquid and slurry materials. Vibrating element technology continues in this use today. At least as far back as the 1960s, vibrating element technology was the basis of level detectors within US patents. Early commercial products are believed to have existed beginning in the 1960–1970s.

The basis of operation for these level detectors is to create a high frequency (perhaps 200–300 Hz) vibration in an element that is inserted into the bin containing the material to be detected. When the vibrating element is free of material it will vibrate at its natural resonance based on the design of the sensor system. When the target material contacts the vibrating element the frequency is dampened and this is detected by the sensor electronics, hence the presence of the material is detected.

There currently exist two primary methods for initializing the vibrating element. The first uses electromagnet elements, and early patents for this technology date back to the 1950s. An excerpt from the abstract of a more recent patent (Myers, 1975) reads as follows:

The vibratory system comprises inner and outer rod elements projecting transversely from the inner and outer sides of the wall member and secured thereto, as by an internal screw stud with the axes of the rod members substantially aligned. Vibrations are induced in the outer rod element by an **electromagnet** *having a coil which is supplied with low frequency electrical pulses. The magnitude of such vibrations is sensed by sensing the electrical signals induced in the coil due to such vibrations. A biasing magnetic flux is produced between the* **electromagnet** *and the outer rod element so that the vibrations produce variations in the magnetic flux. The electrical pulses supplied to the coil are at a substantially lower frequency than the natural vibration frequency of the outer rod element. Such electrical signals may be supplied through an amplifier to a comparator for comparing the level of the amplified signals with a reference level, and causing operation of an output device, according to whether or not the level of the amplified signals exceeds the reference level. The amplifier may be gated so that it is inactive during the vibration inducing pulses, while being active during the intervals between such pulses. When the flowable material in the bin or other receptacle engages the inner rod element, the vibrations of the outer rod element are reduced in magnitude. This reduction causes a change in the state of the output device, which may be in the form of an electrical relay adapted to operate a warning device, or to perform any desired control functions.*

The second method to produce vibration in the vibrating element sensor is the use of piezoelectric crystals. In 1969 a patent was filed that illustrated the use of two vibrating rod mechanisms, currently known as a tuning fork, placed in vibration at its mechanical natural resonance by use of piezoelectric crystals (Endress et al., 1969). The patent states the reason for use of a vibrating element sensor for point level detection applications very clearly: "Apparatus of this type is used, more particularly, where the usual capacitative instruments for measuring the filling level of the tank, cannot be employed because of the low dielectric constant of the material in the tank." This gives us a very good understanding of why vibrating element technology evolved and why it remains in use today as the leading point level technology for both liquid and bulk solid materials, arguably so versus rotary paddle technology for bulk solids.

In addition to the tuning fork configuration using piezoelectric crystal technology, vibrating single element rod units are also very common (see Figure 2-15). Single rod units claim to be more tolerant in applications where the material may tend to cling and pack between the two elements of the fork type of device.

Today there remains a distinct advantage of vibrating element technology for use in detecting the presence and absence of powders and granular materials. These are summarized in Table 2-8.

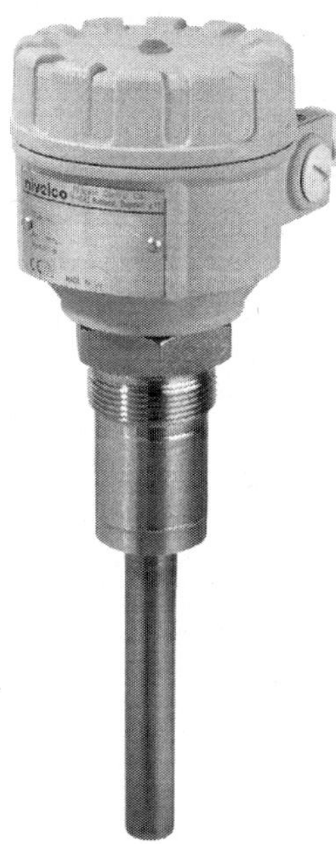

Figure 2-15   Single rod vibrating element point level sensor using piezoelectric crystal technology (Nivocont R brochure, cover, Nivelco Process Control Co., Budapest, Hungary).

Table 2-8   Pros and Cons of Vibrating Element Point Level Sensors

Pros	Cons
Solid-state instrument, no moving parts.	Material buildup can result in false indication of material presence. Use primarily with dry free-flowing materials or those that will naturally shed from the probe element.

(Continued on following page)

Table 2-8   (*Continued*)

Do not require calibration or tuning to the specific application and installation, do not require grounded tank like RF admittance/capacitance.	Typically more expensive than other technologies for bulk solid applications.
Easy to install and set up.	Sensitive to damage by mechanical impact on probe element; has improved in recent years.
Excellent for lightweight materials (as low as 0.624 lb/ft³ or 0.01 kg/dm³) and those with low dielectric constant.	

2.4   Conclusion

The need for point level measurement sensors is widespread across industry. Look around you where you work, live, and play. Many of the things you see that are man made have been touched in some positive way by these level control devices. From plastic parts, to concrete, the shingles on your home or business, food, dog food etc. These items are processed along through their production, most likely in some way aided by the operation of point level sensors. For over a century technology has evolved to perform level control through point level sensing better, more reliably, more accurately, or less expensively than before. The technologies include those which are mechanical, electromechanical and completely solid-state. How they are applied and chosen one over another in specific applications within a wide variety of industries is the subject of Chapter 3.

Study Questions

Q1: What is the primary function of point level sensors? Provide generic examples.

Q2: For powder and granular materials what is arguably the most universal point level sensor technology and why?

Q3: What needs to be considered to properly select the most appropriate technology for a point level application?

Answers

A1:   While exceptions do exist the primary purpose for point level sensors is to provide sensing of material presence or absence in order to control some function. Hence the common term "level control" is used as synonymous with "point level sensor." The most typical examples include the use of a point level sensor as a high level detector to control vessel fill shutoff and prevent overfilling the vessel. Another typical application is that of detecting the absence of material at a low point in a bin to indicate outage so that the material can be reordered or to indicate that a filling process can begin.

A2:   The first technology to be considered as the "king of the bin" (Lewis, 2006) is the rotary paddle point level sensor. The rotary paddle device has very few applications in which it cannot be used. Claims for this type of device exist showing it spans bulk densities from 2 lb/ft^3 (0.3 kg/dm^3) to over 100 lb/ft^3 (1.6 kg/dm^3). Dielectric constant has no direct effect. There is no calibration or tuning. It is tolerant of some buildup and material accumulation. Technically simple and easy to understand, the rotary paddle is very intuitive when it comes to installation, use, and maintenance. One final point that makes rotary paddle technology a king of the bin is its relatively low purchase cost.

In addition to rotary paddle technology the vibrating element point level sensor has become an integral part of bulk solids bin level control. Demand for and usage of vibrating element sensors has grown steadily in the past few decades. The reason for this is that they were first applied primarily as liquid level point level sensor technology and their use with bulk solid materials came later. Success and multiple vendor involvement subsequent to early patents also increased usage of this technology. But the primary reasons for its increased and dominant popularity is similar to that of the rotary paddle; (1) virtually unlimited in material bulk densities, including the ability to sense the presence of materials as lightweight as 0.624 lb/ft^3 (0.01 kg/dm^3), (2) no calibration or tuning is required, and (3) the dielectric properties of the target material have no effect on the operation of the sensor. Vibrating element point level sensor technology is perceived as solid-state with no moving parts. The absence of mechanical movement is not technically accurate since there exists vibration in the sensing element, however, it is a practical claim when compared to the moving parts within a rotary paddle, tilt or diaphragm type level sensor. The only apparent limiting factors are a higher average price structure and sensitivity to material accumulation on the sensor probe.

A3:   While we have not yet reviewed details of application considerations, we have, through our discussion of typical point level sensor applications and the common technologies used for sensors, given glimpses of some of the important parameters. Understanding how each of the technologies used for point level sensing operate we can identify the following application parameters or characteristics that should be considered: (1) bulk density, (2) material agglomeration to the sensing element, (3) material properties such as dielectric constant, (4) particle size, and (5) material flow properties and surface shape or angle of repose of the material surface during fill and discharge of the vessel. These application parameters, along with others, will be discussed further in Chapter 3.

References

Endress, George H. et al. 1969. Apparatus for determining the filling level of a container. US Patent 3,625,058, filed Jul. 7, 1969, and issued Dec. 7, 1971.

Lenhart, Herbert S. 1936. Indicator. US Patent 2,116,075, filed Nov. 28, 1936, and issued May 3, 1938.

Lewis, Joe. 2006. King of the bin: Rotary paddle point level indicator. *Powder and Bulk Engineering* (November).

Myers, Donald M. 1975. Vibratory bin level indicators. US Patent 4,008,613, filed Oct. 3, 1975, and issued Feb. 22, 1977.

Wikipedia. 2010. http://en.wikipedia.org/wiki/Heyl_%26_Patterson_Inc.

Chapter 3

Point Level:
Technology Selection

Objectives

After reviewing this chapter the reader should be able to:

- List and discuss the primary point level application characteristics.
- Understand the impact of the primary application characteristics upon technology choice.

Summary

Chapter 1 identified two types of level measurement devices, that is, point level and continuous level. In Chapter 2 we discussed the use of point level sensors for control purposes and each of the mechanical, electromechanical, and electronic or solid-state devices that are available to detect powder and bulk solid materials at predetermined points. With this foundation we now discuss the application characteristics that impact the selection of the best technology to fit a specific point level application. There are several application characteristics, including temperatures, mounting location, and material properties. Understanding these will allow us to quickly select the best technology for a given application and provide a reference for future use. The choice of a point level monitoring technology is impacted by many factors.

These factors can be categorized into "installation/process" and "material characteristics." They include the following:

- Installation/process
  - Temperature
  - Mounting location
- Material characteristics
  - Bulk density
  - Particle size
  - Dielectric constant (ε_r)
  - Corrosion/abrasion
  - Adhesion

3.1 Installation/Process

Temperature

The temperature of the material to be detected within a vessel, or the ambient environmental temperature surrounding the point level sensor, will in some way impact the sensor's ability to perform reliably, or impact its accuracy in performance. The study of our physical world tells us that materials contract as they become colder and expand as they become warmer. We also know that material moves between solid, liquid, and gaseous states as temperature changes. Therefore, the "temperature effect" on mechanical systems, electronic components, and the environment and world around us impacts our consideration, choice, operation, and the performance of point level sensors. This is an important installation/process characteristic to consider and must be fully understood in order to make the best technology choice the first time.

The "process temperature" is generally considered to be the temperature inside the vessel that contains the bulk solid material or the temperature of the material itself within the vessel. Many bulk solids are processed using temperature. For example, clinker production within the cement production process (Lewis, 2010) uses "pyro-processing"[1] where a kiln is used to bring raw ingredients to a temperature of between 1400°C and 1450°C (2552–2642°F). This high temperature affects the ingredients resulting in chemical reactions that lead to the formation of sintered lumps of the material called clinker. While the clinker will cool relatively quickly from this high temperature, point level

---

[1]http://en.wikipedia.org/wiki/Pyroprocessing

measurement for high level control of clinker material directly out of the kiln can be at a very elevated temperature, well above the local ambient temperature.

During the pelleting process in feed milling the moisture content of the feed mix is increased by the use of steam, which increases the temperature of the mix and pellet from its ambient temperature to around 92°C (198°F) in a matter of about 20 seconds (Hasting and Higgs, 1980). Monitoring the level of material in this mixing process or the level of the pellet within a pellet cooler where the pellet enters at this elevated temperature and cools within a matter of minutes could also impact technology performance and reliability as the temperature, while not extreme, changes rapidly.

Thermal transfer occurs in any one of three ways: conduction, convection, and radiation. Conduction occurs when two items are in contact with one another and heat transfers from the warmer item to the cooler one. At the molecular or particle level the warmer a particle, the faster it moves. As it moves more and more rapidly it expends energy in the form of heat. This heat is transferred to particles adjacent to it, therefore transferring temperature from the warmer particle to the cooler ones. Not all particles or materials transfer heat at the same rate. Metals and stone or rock tend to be good conductors of heat, while things like paper, cloth, and wood are not. Convection is one of the most common methods of heat transfer; however, it does not take place within solids because convection is the movement of molecules and, as we learned in Chapter 1, solid particles do not move at the molecular level. Convection occurs in gases and liquids as the heat is transferred by the object being heated. For example, if the temperature within a bin is very high it can cause the metal of the bin walls to heat or rise in temperature as they are initially at a lower temperature. As the bin walls become hot the heat from the bin wall heats the surrounding air. Air, which is a gas, transfers the heat by movement of air molecules to other items that may be in proximity, such as a measuring instrument or your hand. If you place your hand near the warm or hot surface of the bin wall you may feel the heat rise to your hand. This transfer of heat through the air to your hand is called convection, while the very hot bulk solid material within the bin transfers its heat to the bin wall through conduction. Radiation is the third method of heat transfer. When electromagnetic radiation travels through space it has both electric and magnetic properties and moves in the form of energy with a wave-like motion. Some forms of electromagnetic radiation produce a thermal component: heat generated that flows along with the electromagnetic energy. Two common examples are the sun's light and warmth and the visible light and warmth generated by an incandescent light bulb. Point level sensors can be impacted by their environmental or ambient temperature as well as the material/process temperature. The temperature of the material or process is transferred to the sensor through conduction or convection, while the environmental ambient temperature is transferred to the sensor generally through thermal radiation. Even electronic or solid-state sensors are impacted by temperature. For example, the capacitance of a capacitor can vary with changes in the ambient temperature; similarly, transistors can show an increase in the amount of leakage current or a drop in their rated gain with increase in temperature. Crystals will drift as temperatures fluctuate, and integrated circuit chips may fail if the ambient temperature exceeds their rated range. The ambient and process temperatures also work together to impact the suitability of a product or technology in a given application. While the process temperature—the temperature inside the bin where the sensor is mounted—might be as high as 110°C (230°F) a sensor's rated maximum ambient temperature may be less than this value, perhaps 80°C or 176°F, and cannot be exceeded. The ambient temperature specification must be adhered to or else the sensor may fail to operate properly, if at all. Some manufacturers will provide a chart that describes how the ambient and process or medium temperatures work together in the form of a derating diagram or chart. Refer to Figure 3-1 for an example of a derating diagram.

This derating diagram illustrates that as the process temperature increases the maximum ambient remains constant, until the process temperature reaches its maximum rated level, at which time the maximum allowed ambient temperature begins to decline as the process temperature continues to increase. This type of derating temperature diagram is very helpful to understand the relationship between the process and ambient temperature for a given point level sensor product. It will help ensure the selection of the appropriate device for your application. Refer to Table 3-1 for a general comparison of point level technologies. Some brands may differ.

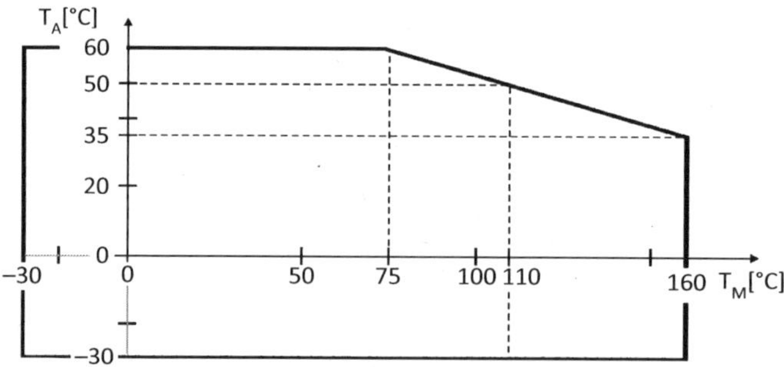

Figure 3-1   Derating temperature diagram for vibrating rod point level sensor (Nivocont R datasheet, Nivelco Process Control Co., Budapest, Hungary).

Table 3-1   Internal Bin and Ambient Temperature Ranges for Common Point Level Sensor Technologies

Technology	Minimum Process Temperature	Maximum Process Temperature	Minimum Ambient Temperature	Maximum Ambient Temperature
Diaphragm	−40°F (−40°C)	250°F (121°C)	−40°F (−40°C)	250°F (121°C)
Tilt	−40°F (−40°C)	250°F (121°C)	−40°F (−40°C)	250°F (121°C)
Rotary paddle	−40°F (−40°C)	300°F (149°C)	−40°F (−40°C)	200°F (93°C)
RF admittance/ capacitance	−40°F (−40°C)	176°F (80°C)	−40°F (−40°C)	150°F (65°C)
Vibrating	−22°F (−30°C)	230°F (110°C)	−22°F (−30°C)	140°F (60°C)
Capacitive proximity switch	−13°F (−25°C)	176°F (80°C)	−13°F (−25°C)	176°F (80°C)

This data is a summarization synthesized from the specifications for a standard product from a variety of different brands. These technologies will be acceptable for at least 90% of the point level monitoring applications in powder and bulk solids. Other selection criteria, in addition to temperature, will be used to make a final decision on which technology will be the best fit.

Mounting Location

The mounting location is very important to consider when selecting a point level sensor. First, the sensing point must be determined based upon the control function required. For example, when choosing a sensing point for a high level control application you should consider the following:

1. Will there be material "run-off" continuing into the vessel even after the high level signal has been activated? Often this is necessary to clear the filling system of material before it shuts down. If this will exist, then the sensing or activation point of the high level sensor must consider the amount of "run-off" that will occur in order to prevent

overfilling or over-pressurization of the vessel as these conditions can have expensive and dangerous repercussions.

2. How quick a response time in sensing presence or absence of material is required of the level control sensor? This relates to the speed of filling or discharging and sensor technologies differ in their response time or activation/deactivation time from milliseconds to several seconds. If filling occurs very quickly, as in small vessels or those with high capacity conveying systems, a technology with a rapid response to material presence and absence should be chosen. In addition, several technologies will provide an adjustable time delay, which needs to be properly set to meet the needs of the specific application. Make sure the minimum and maximum adjustment of time delay will meet your needs before making a final technology selection.

3. Is the choice of mounting location limited due to physical obstructions surrounding the vessel or within the vessel? This may necessitate a technology which has a low invasive profile, is flush mounted to the inside wall of the vessel, or perhaps is very compact so that the sensor exposed on the outside of the vessel will allow for easy installation and maintenance. Perhaps top mounting is the only choice, but there is very limited headroom to insert or remove a sensor? This will necessitate a low profile, compact, or flexible device.

Technology and product choice for the best point level sensor for any given application can involve ensuring the physical size of the sensor is acceptable, looking at accessibility requirements to the sensor, material flow within the bin, and how any equipment surrounding the vessel may limit access to the sensor. How does material flow impact the mounting location or sensing point?

The location of the material inlet and discharge within the vessel, as well as the material flow characteristics, will determine the angle of repose. Based upon your level control objective and the shape of the material surface during filling and discharge, this can help lead you to the optimal sensing point for either high or low level. As an example (see Figure 3-2), in a high level control application for material storage you may want to maximize the amount of material within the vessel. If the vessel has an off-center fill inlet, and a center discharge point, the angle of repose could be at a steep angle, limiting the amount of material that can be stored and dictating your high level sensing point accordingly.

One final point regarding mounting location is with regard to other devices mounted on the bin in close proximity to where you want to mount the point level sensor. Specifically, the existence of an industrial vibrator, viable for

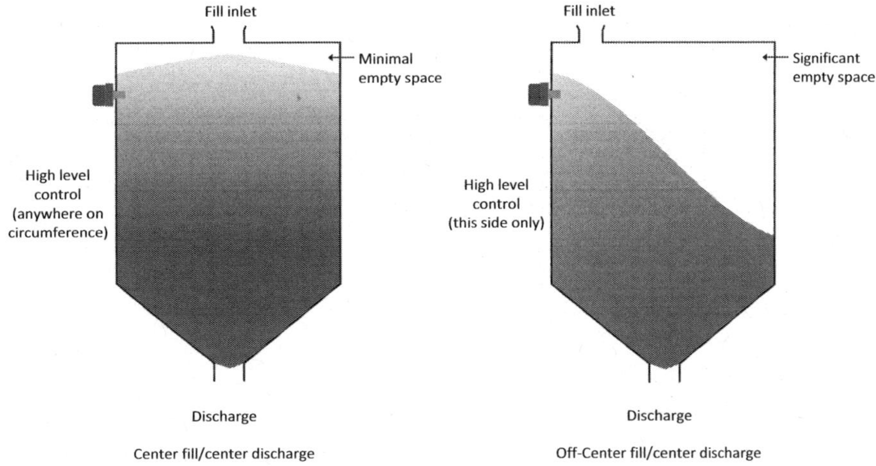

Figure 3-2   Angle of repose can impact sensing point for level control.

fluidizing the material in the bin to get the material to flow, can introduce high intensity energy onto the bin that can be very destructive to sensors mounted close by. This is especially true for electronic solid-state level sensors such as RF admittance/capacitance sensors and vibrating element sensors (rods or forks). However, mechanical and electromechanical sensors can also be damaged if they are located too close to an industrial vibrator on the bin wall.

3.2   Material Characteristics

Bulk Density

When we speak of density (Cocco, 2010) with regard to solid particles, its definition is dependent upon what density we are speaking of. It seems there are several common "densities" so we need to be specific. These include "absolute powder density," "particle density," "envelope density," "skeletal density," "bulk density," and "emulsion density." For the purposes of using density with regard to any level control or level measurement based inventory monitoring application we use the more common term of bulk density. Bulk density is related to the ratio of the mass of a bulk solid within a vessel to the volume displaced within the vessel by the bulk solid material. This density value includes all voids, both within the particles and between the particles. However, bulk density can be different depending on whether a loose or packed quantity of material is being weighed. Bulk density is usually determined by filling a container with a known volume of empty space with the subject material and then weighing it. However, tapping the container or tamping it so that the material packs will allow more material to fit in the container of fixed volume, thereby giving you a greater bulk density weight than if the material loosely filled the container. Therefore, there exist two types of bulk densities, loose bulk density and packed or tapped bulk density. It is important to understand what bulk density is used or if you are just given the range of bulk density between loose and packed.

No matter how you define bulk density, why does it matter when it comes to point level sensor control applications? The greater the bulk density, the more probability there is for physical damage to invasive sensor technologies that may have a limited ability to withstand the side loading that results from shifting or flowing material, or that can be damaged by falling material if the sensor element is not protected (see Figure 3-3).

The amount of side loading or the force with which material contacts the sensor element is dependent upon the bulk density of the material. As an example, vibrating element level sensors can be damaged by falling material and heavy side loading. This is because this technology creates a vibration in the mechanical structure of the sensor probe at the mechanical system's natural resonant frequency. Even slight bending of the mechanical system can impact operation and performance. Most vibrating element point level sensors in the market today are very robust, with the ability to withstand side load forces up to 500 N (newtons), torque forces of up to 100 Nm (newton-meter) and pull down tensile forces of up to 100 Nm. Let us be clear about what this means. A newton is a measurement unit for force. One newton is the amount of force required to accelerate a mass of 1 kg (2.2 lb) at a rate of

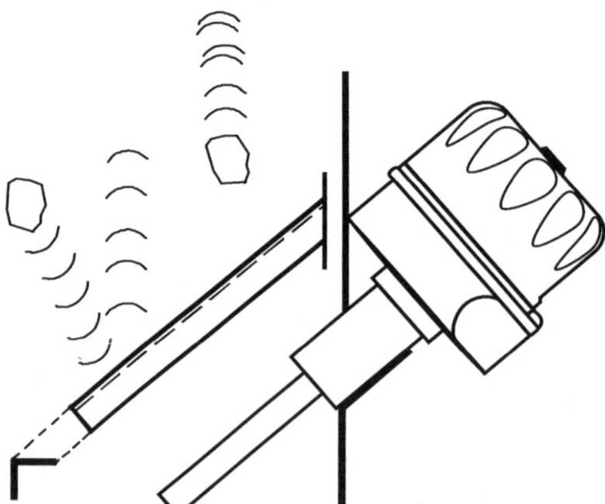

Figure 3-3   Protective plate over vibrating element point level sensor being used for low level control.

1 meter per second per second. This loading from the side or pulling down equates to about 112 lb of force on the probe. Torque is the tendency of force to rotate an object. The torque of 100 Nm is equivalent to about 74 foot-lb of torque force. While the manufacture and design of vibrating element sensors has improved over the last 15 years, this technology should still be avoided in low level applications when the material bulk density is greater than 45 lb/ft^3 (0.72 g/cm^3), unless adequate precautions are taken during installation to protect the sensor probe. Bulk density can also present problems due to the difference between the loose density and packed density. Material at the top of the pile within a vessel is loose in comparison with the density of the material at the bottom of the pile. This means that the bulk density actually changes within the pile of material, and with very lightweight material this change can be significant. These very low bulk densities may be problematic for many technologies so that one device may work for low level, but not for high level. Typical examples include: (1) the need to use a different paddle attachment for a rotary paddle device in a high level application than you do for the low level and (2) the calibration of an RF admittance/capacitance device for a low dielectric material in a low level control position but due to the low dielectric and less packing the unit cannot be properly be calibrated in the high level application position.

Particle Size

Particle sizes for typical bulk solids can range from less than 1 micron to a size measured in inches. Books have been written on the subject of particle size measurement, reduction, and analysis (Allen, 1997). An entire industry exists and is dedicated to its theory and practice. Searching through the online encyclopedia *Wikipedia* yields a mind-numbing amount of technical information on the differences between a powder, granular material, etc. For point level measurement we simply need to understand how the size of a bulk solid particle can affect either the choice of the most appropriate technology for a given application or the technology's performance in the application. The most common concern is the contact of the particle with the invasive part of the sensor element. This involves two different aspects of the contact, the amount of surface contact and the type of contact.

The larger the particle, the more impact the material can have on the portion of the point level sensor exposed to the material inside the vessel. Imagine falling or shifting rock and aggregate material with larger rock-sized particles. Heavy powders will often flow around the sensor probe, while larger particles will make more direct impact. This direct impact can damage some sensor technologies.

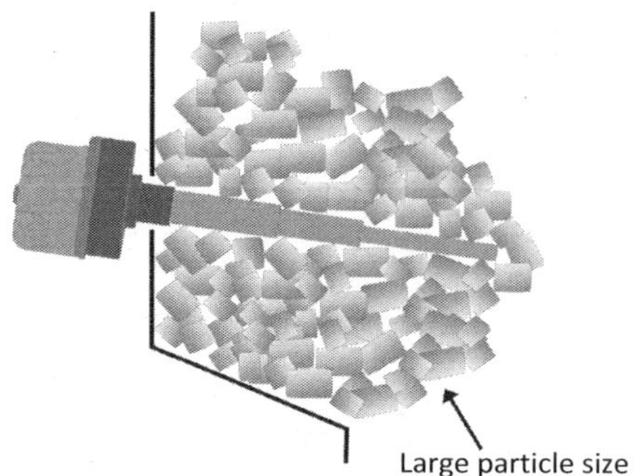

Figure 3-4   Air gaps between particles may reduce the dielectric seen by RF admittance probes below minimum threshold.

At the very least, protective plates or baffles may need to be installed to protect the sensors from direct impact from the material (Figure 3-3).

In addition, the larger the particle size the less potential surface of the sensor probe the material may actually be in contact with. This is due to the space created between particles when the particle size is very large (rock-like). A very small amount of physical contact between the material and the sensor may actually occur (see Figure 3-4).

The space between particles is air, which has the lowest dielectric constant (1.0) of all materials. Enough air gaps and this could be a problem for RF admittance or capacitance point level sensors, depending on the dielectric constant of the material itself, as they rely on the change in dielectric in the area surrounding the sensor probe in order to detect the presence or absence of the material. At the opposite end of the spectrum are materials with very fine micron or sub-micron size particles. These materials are often conveyed using pneumatic systems and the material enters the vessel as a cloud of "dust" being a combination of air and fine particulate. The material bulk density with these types of fine materials can vary significantly within the pile of material from the top to the bottom. The air escapes as the powder settles. For a fine particulate in a high level condition, like back-up protection in the hopper of a dust collector, the dust will be highly aerated and unpacked and this must be taken into consideration when selecting the point level sensor technology.

Dielectric Constant

The dielectric constant of a material is also known as relative permittivity and is closely related to the definition of capacitance. This is a hint of how the dielectric constant of a material that needs to be sensed impacts the choice or performance of point level technology. But first, let us discuss exactly what the dielectric constant[2] is.

The dielectric constant of a material is also known as the relative permittivity of the material, which is the ratio of the permittivity of the material to the permittivity of air in a vacuum. The latter is always 1.0. Therefore, the dielectric constant of any material is always greater than 1.0. But what is permittivity? Permittivity (ε) is a measure of a material's ability to be polarized by an electric charge. I like to think of it as how good an electrical insulator a material is. Permittivity is measured in terms of farads per meter, and is dimensionless. This means that it is just a pure number. The higher the permittivity of a material is, the greater the ability of that material to be a conductor of electricity. The lower the permittivity of the material is, the greater the ability to be an insulator of electricity. Table 3-2 provides a list of common materials and their dielectric constants (ε_r). Many other materials, bulk solids and liquids, exist and dielectric constants can vary by sample, test method, temperature, etc.

So now we know that the dielectric constant of a material, or the material's relative permittivity, is a statement of how well the material behaves as an insulator or conductor of electricity. But why is that important in consideration of the selection of a point level sensor or the sensor performance? This issue relates specifically to the RF admittance/capacitance type sensors.

As discussed in Chapter 2, the RF admittance/capacitance point level sensor operates by being able to detect a shift in electrical admittance or capacitance due to the presence of the target material surrounding and in contact with the sensor probe. The basic formula for capacitance can be written as $C = kA/d$, where k is the dielectric constant of the material to be sensed. The lower the dielectric constant, the smaller the change in capacitance the sensor electronics will need to be able to detect. During sensor calibration the sensor is "tuned" to the specific installation and material. This tuning adjusts the sensor's sensitivity, typically expressed in terms of a minimum capacitance that can be sensed such as 2.0 pf (picofarad), 1.0 pf, or 0.5 pf. For these sensors we need to consider the dielectric constant of the target material to determine whether they are a good choice for the material and application based on the sensitivity of the specific sensor brand, as this can vary. In addition, because these sensors operate on the capacitance sensing principle they generally require a good electric ground at the mounting point and therefore are not considered suitable for non-metallic vessels.

---

[2]http://www.siliconfareast.com/dielectric-constant.htm

Table 3-2   Dielectric Constants of Common Materials

Material	Dielectric Constant
ABS resin, pellet	1.5–2.5
Acrylic resin	2.7–4.5
Aluminum powder	1.6–1.8
Asbestos	3–3.5
Ash (fly ash)	1.7–2.0
Asphalt (at 75°F)	2.6
Asphalt, liquid	2.5–3.2
Bakelite	3.5–5.0
Beeswax	2.7–3.0
Calcium carbonate	6.1–9.1
Carbon black	2.5–3.0
Cellulose	3.2–7.5
Cement, Portland	2.5–2.6
Cereals, dry	3.0–5.0
Charcoal	1.2–1.8
Clay	1.8–2.8
Cocaine	3.1
Corn	5.0–10.0
Cotton	1.3–1.4
Ebonite	2.5–2.9
Ethanol (at 77°F)	24.3
Ferromanganese	5.0–5.2
Glass, bead	3.1
Gypsum	2.5–6.0
Jet fuel (JP4; at 70°F)	1.7
Lead carbonate	18.1
Lime	2.2–2.5
Magnesium oxide	9.7
Neoprene	6.0–9.0
Nylon resin	3.0–5.0
Paper	2.0

(Continued on following page)

Table 3-2   (*Continued*)

Paraffin wax	2.1–2.5
Perlite	1.3–1.4
Plaster	2.5–6.0
Polycarbonate	2.9–3.0
Polyester resin	2.8–4.5
Polyethylene, pellet	1.5
Polypropylene	1.5
Porcelain	5.0–7.0
Potassium chloride	4.6
PVC, powder	1.4
Quartz	4.2–4.4
Refractory (for casting)	1.8–2.1
Rice bran	1.4–2.0
Rubber	3.0
Salt	3.0–15.0
Sand	3.0–5.0
Silica sand	2.5–3.5
Silk	2.5–3.5
Slaked lime, powder	2.0–3.5
Soy beans	2.8
Styrene	2.3–3.4
Sugar, granulated	1.5–2.2
Teflon, PTFE	2.0
Tobacco	1.6–1.7
Urethane resin	6.5–7.1
Vinyl chloride (acetate)	3.0–3.1
Water (at 68°F)	80.4
Wood, dry	2.0–6.0
Zinc sulfide	8.2

Source: Flowmeter Directory, http://www.flowmeterdirectory.com/
dielectric_constant_01.html

Corrosion

Corrosion is defined as the degradation of a material due to exposure to its environmental surroundings (Shaw and Kelly, 2006). There is no way to prevent it—like death and taxes it is unavoidable. What we need to be concerned with for point level measurement is carefully matching the sensor materials of construction with the sensor environment, especially with the material to be detected when the sensor is invasive and will be in contact with the target material.

How does corrosion work? Studies have shown that the annual cost of corrosion is approximately $276 billion per year! That's a staggering amount. Corrosion is an electrochemical process. Even aluminum, stainless steel, and other metals that are more resistant to corrosion than carbon steel will degrade in some manner over time; temperature and the exposure time to certain materials may hasten the process of corrosion. In most applications the use of galvanized or stainless steel materials of construction in the level sensor will protect against uniform or common corrosion.

A subject closely related to and lumped in with corrosion by many people is that of the chemical compatibility between the materials of construction of the level sensor with the environment and target material to be sensed. Chemical compatibility is an indication of how stable a material will be when exposed to another material or chemical. This applies to metals, plastics, and elastomeric materials. Most level sensors contain a combination of metal, plastic, and elastomeric materials. For example, consider ammonium phosphate. This is a chemical compound used in agriculture in some fertilizers, as it is a source of nitrogen. Examining a chemical compatibility chart[3] we find that while stainless steel is rated as "Fair," to be used under special conditions, titanium and certain plastics such as polypropylene and Teflon are rated as "Excellent." In addition, when we examine compatibility with seal materials found in level sensors, such as NBR (Nitril butadiene rubber), EPDM (ethylene propylene diene monomer), and FKM (fluoroelastomer), we learn that all are acceptable. Caution: Many manufacturers publish chemical compatibility charts. Be careful to examine the source of their data and recommendations. Reference tools that should be considered for examination of corrosion and compatibility of a wide variety of materials are the *NACE Corrosion Engineer's Reference Book*, which is currently in its third edition, and the *ASM Handbook Volume 13B, Corrosion: Materials*. In addition, Table 3-3 provides a list of typical materials of construction for standard point level sensor technologies. Consult your chosen level sensor manufacturer or their documentation

---

[3]http://www.flowline.com/chemical-compatibility.php

Table 3-3   Typical Wetted Materials of Construction for Common Point Level Sensor Technologies

Sensor Technology	Common Wetted Materials of Construction
Diaphragm level switch	Neoprene, Teflon, stainless steel
Tilt switch	Aluminum, nickel plated steel, painted steel, plastic, stainless steel
Rotary paddle	Aluminum, powder coated aluminum, stainless steel, NBR, FKM, Viton, plastic
RF admittance/capacitance	Aluminum, powder coated aluminum, stainless steel, Ryton, nylon, NBR
Capacitive proximity switch	Plastics
Vibrating element	Stainless steel

for correct information regarding their specific materials of construction and corrosion resistance or chemical compatibility.

Abrasion

Abrasion is the wear of one material on another. The wear is the erosion of material from a surface by contact with the surface of another material. Affecting the amount of abrasion or wear includes the type of contact between the materials, whether the contact is an impact, a sliding motion, a reciprocating motion, etc. It also depends upon the relative hardness of the surface of each material and the force and speed of the contact. Therefore, in low level applications using invasive point level technologies, the weight of the material on the level sensor probe, the speed of its flow by the probe as the material discharges and the relative hardness of the material versus the level sensor probe material should be considered. For example, the use of a diaphragm switch with a neoprene diaphragm in a high-flow low level application with whole grain like corn may mean a limited life expectancy for the level sensor due to abrasion. A stainless steel diaphragm or a different technology using stainless steel wetted components may be a better choice.

Unfortunately, unlike corrosion and chemical compatibility, there is very little documented reference material available that might provide relative abrasion ratings for materials other than metals. It is best to thoroughly understand your target material and the dynamics of the application where the level control sensor is to be installed, and to consult with your chosen level sensor supplier.

Adhesion

The ability of unlike substances to stick to one another is known as adhesion. Cohesion, a related term, is the ability of like substances or molecules to stick together. It is possible for the particles of a material to stick together and have strong cohesion, while they have relatively weak adhesive properties with most common materials. While contact of the material to be sensed with the level sensor probe element is usually necessary and desirable, the adhesion of the target material to the sensing element is not. Too much material adhesion can result in false indication of material presence. In some technologies even just a little adhesion may be problematic. Table 3-4 summarizes the sensitivity of the various technologies to adhesion of material.

How do you know whether a material will adhere to the level sensor element in enough magnitude to create a sensing problem? One way is to examine the material in its container if it already exists. If the material clings and adheres to the vessel straight walls, then it may do likewise with the level sensor. Choosing a level sensor that has low sensitivity to material buildup and coating, or one that has the ability to be immune to the effects of coating or buildup, will be important. Does the material flow in the straight wall vertical section but get stuck in the cone area or the transition between the straight wall and the cone? This indicates that the material flows somewhat well enough, but that gravitational forces are not strong enough to break the adhesive bonds between the material and the vessel cone walls, resulting in a possible adhesion problem for the level sensor, perhaps depending upon installation location. This suggests that choosing a level sensor technology that has low to moderate sensitivity to material buildup would be best. No formula exists to grade the adhesive properties of various bulk solid materials with the materials of construction of the possible level sensor technologies that could be used for a given application. Knowledge of the target material is the key. If you do not have any idea about this, check it out by examining the material itself or as it exists and flows within the vessel that contains it. Check the material safety data sheet (MSDS) before attempting the following empirical test. Use appropriate protective gloves. Take a handful of the material and squeeze it together in your hand. Does the material clump together? If yes, then the material has cohesive properties and may be adhesive in nature as well. If not, it is unlikely to be adhesive enough to cause

Table 3-4   Sensitivity of Level Sensor Technologies to Material Adhesion/Coating/Buildup

Level Sensor Technology	Sensitivity to Material Adhesion
Diaphragm switch	*High sensitivity* Should be used with dry free-flowing materials; sticking material on diaphragm may activate switch and falsely indicate material presence, especially prevalent with side-mounted and low level applications with adhesive materials; consult manufacturer for guidance
Tilt switch	*Very low sensitivity* Not a problem since these switches are used for high level only and require significant material presence buildup from beneath the sensor to tilt it by 15–17° for activation
Rotary paddle	*Low–moderate sensitivity* Sensitivity is dependent on the magnitude of the buildup; some adhesive materials can ingress through the sensor seal and become "gummy," reducing effective operation or causing the level sensor to malfunction; consult manufacturer for guidance
RF admittance/capacitance	*High* *Note*: Low–moderate sensitivity to adhesive materials if level sensor is equipped with automatic buildup immunity feature such as a guard or shield section; in this case the sensitivity will be dependent on the amount of buildup and the dielectric constant of the buildup material; consult manufacturer for guidance

(*Continued on following page*)

Table 3-4　(Continued)

Capacitive proximity switch	*High sensitivity* These devices must be used with dry free-flowing powder and granular materials or clean non-adhesive liquids such as water; these units are not tolerant of adhesive buildup, unless the buildup is non-conductive very low dielectric constant material, which would be very difficult to sense properly if it were the target material
Vibrating element	*Moderate–high sensitivity* Not tolerant of adhesive buildup; dusting or very light buildup may be okay if sensor can detect target material in high density setting; consult manufacturer for guidance

a buildup problem for most level sensor technologies. If the material clumps in your hand, throw a clumped handful against a wall. If some of the material sticks to the wall, then it has relatively good adhesive properties and a level sensor technology with low sensitivity to adhesion or with buildup immunity should be considered. In addition, you should consult with your chosen level sensor supplier for their collective experience in this area. However, in the end it is the responsibility of the material manufacturer or user of the level sensor to understand the adhesive properties and nature of the material.

3.3　Conclusion

Point level monitoring or level control is a critically important application. High level detection is used to control a vessel filling process and prevent costly and dangerous overfilling. Low level monitoring and control can ensure timely notification that reordering of material is required to avoid material outages and a production shutdown. Several viable technology choices exist and these were all reviewed in Chapter 2. Several application and material characteristics exist that will need to be considered before deciding which level sensor technology may be the best device for the application. Sometimes a clear superior technology is evident. At other times no clear "best choice" may exist or there may be

Table 3-5   Technology Selection Guide based on Common Application Characteristics

	Diaphragm	Tilt	Rotary	RF	Vibrating	Prox
Material						
Powder	Y	Y	Y	Y	Y	Y
Granular	Y	Y	Y	Y	Y	Y
Slurry	N	N	N	Y	CF	N
Liquid	N	N	CF	Y	CF	Y
Bulk Density						
Very low 1.5–5 lb/ft^3 (24–80 kg/m^3)	N	N	CF	CF	Y	CF
Low >5 lb/ft^3 (>80 kg/m^3)	CF	CF	Y	CF	Y	CF
Medium >15 lb/ft^3 (>240.3 kg/m^3)	Y	Y	Y	Y	Y	Y
High >40 lb/ft^3 (>640.7 kg/m^3)	Y	Y	Y	Y	CF	Y
Moisture						
Low	Y	Y	Y	Y	Y	Y
High	Y	Y	Y	Y	Y	Y

(Continued on following page)

Table 3-5   (*Continued*)

Process Temperature						
>200°F (93°C)	Y	Y	Y	CF	CF	N
>300°F (150°C)	CF	N	CF	CF	CF	N
Material Coating						
Minimal	Y	Y	Y	Y	CF	CF
Heavy	N	CF	CF	Y	N	N

Y = Yes; N = No; CF = Consult factory.

multiple possibilities. Table 3-5 is a summary of these choices and is provided as a guide and recommendation. This table will also indicate which of these sensor technologies may be suitable for slurry and liquids. A slurry material is a viscous liquid with a high content of insoluble solids. Viscosity varies.

Study Questions

Q1: What are the primary application characteristic categories for point level sensors? Provide some examples.

Q2: Discuss the issues to consider if you are contemplating the use of RF admittance or capacitance point level sensor technology.

Q3: Is there anything else that may impact the point level sensor we choose for a given application that we have not yet discussed?

Answers

A1: There are several important application characteristics or parameters that need to be considered to determine the best choices available for any given point level application. These can be categorized into installation/process and material characteristics. In installation/process this includes temperature and mounting location. Temperature factors include the process or internal temperature within the vessel or the target material temperature, as well as the ambient temperature surrounding the enclosure of the level sensor on the outside of the vessel. With the mounting location we may need to consider how large the point level sensor is and how much room exists for sensor insertion, installation and operation. This may require a very compact sensor or may not impact the choice of sensor technology at all, but it needs to be considered. Material characteristics requiring consideration are several, including bulk density, particle size, dielectric constant, corrosion, abrasion, and adhesion. All stand on their own being able to exclude or include a particular type of level sensor, however, adhesion stands out as being an issue for all technologies. The magnitude of the adhesive capability of a material can possibly rule out all technologies discussed thus far.

A2: RF admittance/capacitance technology exists to offer a solid-state alternative to rotary paddle and other mechanical or electromechanical technologies in the hope of less maintenance and longer life. However, these devices are directly impacted and limited by the dielectric constant property of the target material and the need for installation in a metal vessel (or a plastic vessel with a good ground connected to the sensor external ground connection on the enclosure). These are the issues, in addition to other general issues, that specifically apply to RF admittance/capacitance point level sensors. Because of their principle of operation, materials with low and changing dielectric constants present a challenge requiring very high sensitivity in terms of picofarads. Typically the highest sensitivity available will allow the RF unit to detect as small as 0.3 pf change in capacitance. RF admittance/capacitance point level sensors require calibration or tuning to the specific application. This is an extra step in the installation and setup of the device that does not exist in other technologies. In addition, if the dielectric constant of the material changes appreciably (up or down), recalibration may be required. The dielectric constant can change with moisture content or a change in the target material within the vessel.

A3:   Yes. The electrical classification of the location where the point level sensor will be installed needs to be considered. This is a safety issue for the facility and personnel. Generally speaking there are two types of locations; those considered ordinary or general-purpose industrial environments, and those considered hazardous due to the presence of explosive vapor or dust. This does not necessarily affect the choice of technology to be used (the reason it has not been mentioned in this chapter), but it does impact the version of the device you might choose, the brand you may choose, as well as possibly impacting the cost of the device and its installation.

References

Allen, Terrence. 1997. *Particle size measurement*. Powder Technology Series, vol. 1. 5th ed. London: Chapman & Hall.

Cocco, Ray. 2010. What is density in particle technology? *Powder and Bulk Engineering 24* (October): 16–21.

Hasting, W. H., and D. Higgs. 1980. Feed milling process. In *Fish feed technology. Aquaculture Development and Coordination Programmme*, Chapter 18. Rome: Food and Agricultural Organization of the United Nations. http://www.fao.org/docrep/x5738e/x5738e0j.htm

Lewis, Joe. 2010. Bin level indication applications in cement production and concrete batching plants. BlueLevel Technologies, Inc., Rock Falls, Illinois.

Shaw, Barbara A., and Robert G. Kelly. 2006. What is corrosion? *The Electrochemical Society Interface*, Spring, 24–26.

Chapter 4

Point Level: Specialty Technologies

Objectives

After reviewing this chapter the reader should be able to:

- Identify specialty technologies for point level measurement.
- Understand the use and primary application of these specialty technologies.

Summary

Chapter 2 identified and discussed the most commonly used types of technologies for point level measurement and control, that is, pressure-sensitive diaphragm switch, tilt switch, rotary paddle, capacitive proximity sensors, RF admittance/capacitance and vibrating element. In Chapter 3 we discussed the characteristics of the mounting location and material that impact our choice of technology. But there are still some applications where point level measurement and control are required, but the commonly used technologies of Chapter 2 are not the best fit, if at all. This includes applications where environmental conditions or material characteristics totally exclude the use of point level technologies that are invasive to the vessel and in contact with the target material. Two additional technologies need to be mentioned and discussed, along with their typical use and limitations if any. These specialty technologies include beam-breaker and radiometric devices. Both are non-invasive and, at least potentially, non-contact in use for point level measurement and control.

With the foundation of Chapters 2 and 3, we will review beam-breaker and radiometric technologies and discuss their application and use, including the characteristics that make them the best or only choice available. As before, we will also include a table of the pros and cons for these devices.

4.1   Beam-Breaker

The challenges of some applications necessitate a minimally invasive or non-contact point level sensor technology. Coal, rock, and other materials with large and heavy particles can damage sensors with invasive probes because they are heavy and abrasive. Other materials may be so abrasive or corrosive that a non-contact device is the only possibility. The beam-breaker type of point level sensor offers a non-invasive installation, and in some situations where a non-conductive window can be installed or a non-conductive vessel is used, this technology can be non-contact as well.

The most common type of beam-breaker device for bulk solids is based on microwave technology. There are light-based or optical technologies, however, these units are so sensitive to dust that they are impractical in bulk solid applications. Most microwave based beam-breaker point level sensors use an X-band 10 GHz pulsed microwave energy signal. However, some also use the K-band, which operates at 24 GHz. All beam-breakers consist of two primary components, the transmitter and the receiver. These components are housed in two separate enclosures with signal wiring connected to the central processor in its enclosure. A typical microwave beam-breaker point level sensor are shown in Figures 4-1 and 4-2.

Microwave beam-breaker point level instruments generate the microwave pulse signal in the transmitter. The transmitter is installed on one side of the vessel and the receiver on the other. Because the antennae are small, the beam angle of these devices is relatively large, for example, 20–40°, depending on the frequency, and therefore alignment of the transmitter and receiver does not have to be perfect. Microwave output power is usually less than 10 mW and therefore complies with Federal Communications Commission (FCC) Title Rule 15 and Occupational Safety and Health Administration (OSHA) exposure specifications in the United States. Note that FCC Title 15 sets out the conditions, technical specification, and circumstances whereby the intentional generation of radiated energy, such as in a microwave beam-breaker

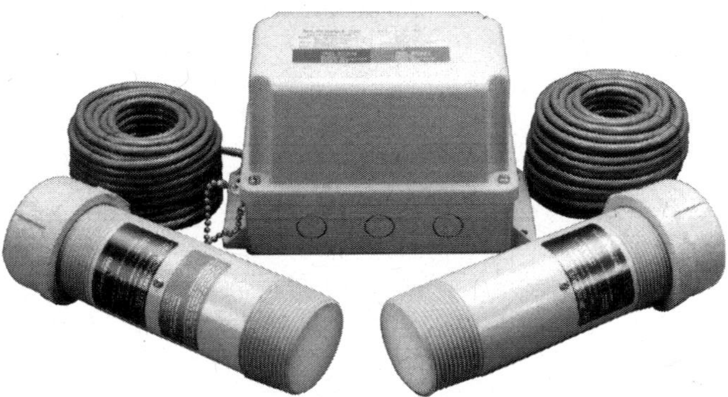

Figure 4-1 X-Band 10 GHz microwave beam-breaker point level sensor (Microwave 320, front page data sheet, Position & Level Control, PDS-D320-1B, Delavan Process Instrumentation, An L&J Technologies Company, Hillside, Illinois).

Figure 4-2 K-Band 24 GHz microwave beam-breaker (Model MWS-ST/SR-2, Microwave Level Switch Operation Manual front cover, WADECO Wire Automatic Device Company, Ltd., Hyogo-Ken, Japan).

device, can be done without an FCC license being required. In addition, as for safety the OSHA defines the frequency limits for radio frequency energy as being between 3 kHz and 300 kHz, while the frequency limits for microwave radiated energy is between 300 MHz and 300 GHz. OSHA specification 29

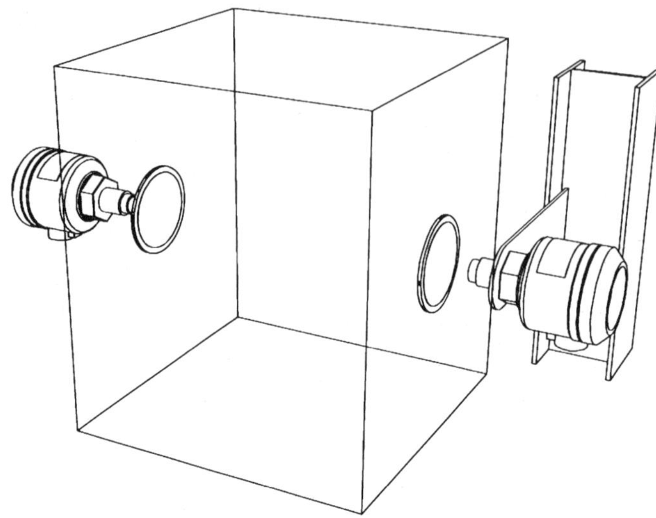

Figure 4-3   Microwave beam-breaker point level sensor installed with non-conductive windows (Model MWS-ST/SR-2 Series Level Switch Operation Manual, page 9, WADECO Wire Automation Device, Co., Ltd, Hyogo-Ken, Japan).

CFR 1910 subsection 1910.97 limits exposure of electromagnetic radiation from 10 MHz to 100 GHz to a power of 10 mW/cm^2 over a 0.1 hour time period, whether continuous or intermittent.

The microwave beam-breaker transmitter emits a microwave energy beam across the measuring point area within the vessel as shown in Figure 4-3. When no material is present within the measuring area (free air) a strong level of microwave energy from the transmitter will be detected at the receiver component. When material level enters the measuring area the microwave energy detected at the receiver will diminish as a result of energy being reflected by the material. A sensitivity adjustment is provided to allow the sensor to be "tuned" to the specific detection requirement of the application. The control output from microwave beam-breaker point level sensors is like that of other point level sensor devices, relay contacts. Some units offer an optional analog output signal. These devices should not be confused with continuous level measurement sensors, to be discussed in a later chapter. Operating temperatures are generally at normal ambient conditions ranging from −40°F to +158°F (−40°C to +70°C). Some brands may offer transmitter/receivers that can operate in environments where higher temperatures exist, perhaps up to around 1,100°F (600°C).

Table 4-1 Pros and Cons of Microwave Beam-Breaker
Point Level Sensors

Pros	Cons
Non-invasive in virtually all installations.	Higher purchase cost than most all point level sensors.
Non-contact with material when installed in non-conductive vessel or with a non-conductive window.	Performance dependent on dielectric constant of target material.
Minimal impact from abrasive and large particle size material.	Coating or material buildup of materials with high dielectric constants can produce false signals. Not so much with non-conductive low dielectric material buildup.
In most situations the presence of dust during filling has minimal impact on performance of the sensor.	

As the energy beam angle is relatively large, the installation of the transmitter and receiver components does not have to be perfectly at the same height within the vessel. However, care should be taken to align the components as best as possible opposite to each other. Microwave beam-breakers are generally installed so that the components are non-invasive to the material within the vessel. In addition, because microwave energy can penetrate materials with low dielectric constants, such as refractory material, plastic, ceramic, glass, the transmitter and receiver components can be installed so that they will not be in contact with the material within the vessel by using these materials as the vessel wall or by installing a window in the vessel wall. Refer to Table 4-1 for a list of Pros and Cons of microwave beam-breaker point level sensors.

4.2   Radiometric Point Level Sensor

Also known as radiation-based or nuclear level sensors, these devices are used in select applications where extreme conditions make them the only choice. These conditions include temperature and pressure extremes, aggressive materials that are highly abrasive or corrosive, along with extremely toxic or carcinogenic materials whose containment cannot be breached. The principle use of radiation-based level monitors is where non-contact and non-invasive sensor technology is a must.

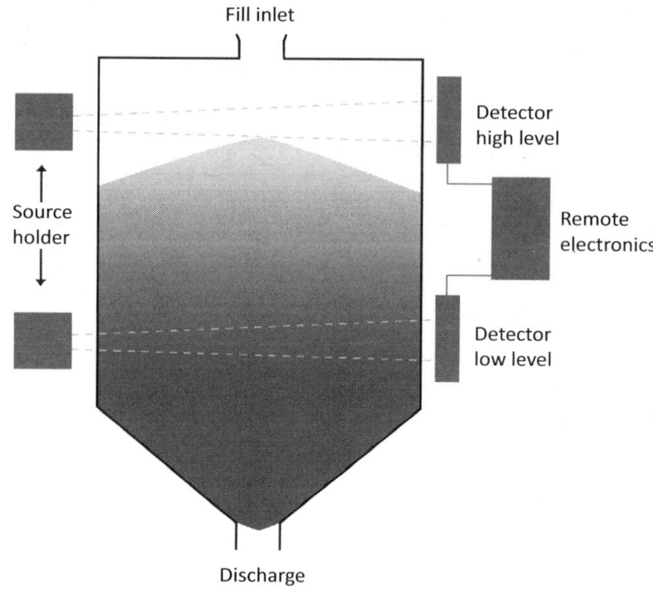

Figure 4-4　Radiation-based point level sensor for a bulk solid.

Radiometric level sensors consist of three primary components: the source holder, the detector, and the sensor electronics. The electronics may be remotely mounted from the source holder and detector or included within the detector assembly. The source holder will contain the source material that emits gamma energy, which is directed by the source holder. The source holder is mounted on an outside wall of the vessel containing the material to be detected. The source holder and material are never in contact with the material in the vessel. Opposite the source holder, mounted on the outside of the vessel wall, is the detector assembly. During the absence of material inside the vessel between the source holder and the detector, the gamma energy emitted from the source holder travels through the vessel wall and into the atmosphere within the vessel and then passes through the opposite vessel wall where it is sensed by the detector. The vessel walls will reduce the gamma energy. However, the detector has the necessary sensitivity to detect the lower energy levels. As material displaces the internal atmosphere and enters the area between the source holder and the detector, the gamma energy decreases again and the detector senses this and determines that material is present. Figure 4-4 illustrates a point level sensing example using radiation-based devices for both high and low level.

Detector Technology

An important aspect of radiation-based point level sensors is the detector technology (Liptak et al., 2003). There are three primary methods for detecting the gamma energy emitted from the source in radiation-based point level sensors, that is, GM tubes, ionization chambers, and scintillation detectors. Most common in point level sensors for years the GM tubes, or Geiger Müller tubes, consist of two components, the anode and the cathode. GM tubes are also the sensing element used in Geiger counters that detect ionizing radiation, such as gamma energy, which is ionizing due to its high-energy photons. A high voltage potential of several hundred volts will exist between the anode and cathode without any current flow. When gamma energy passes through the tube, some of the inert gas (such as helium) is ionized. This ionization of gas particles increases and a current flows between the two nodes from negative to positive. This is analyzed by the sensor electronics, which in turn distinguishes between the absence and presence of material within the vessel.

Ionization chambers operate similarly to the GM tubes except the voltage used is very low, in the range of several volts, and the resultant current due to ionization of the gas particles by the presence of gamma energy is much smaller, in the microampere range. This makes managing any leakage currents important. However, ionization chambers are not as common as the GM tubes or scintillation detectors.

Scintillation has become popular due to its improvement in sensitivity with a resultant decrease in source material yielding improved measurement capability with greater safety and lower cost. The term "scintillation" is based upon the use of a material that will emit light in response to ionization by a radiation source such as gamma energy. In operation the scintillating material will increasingly emit light as a result of an increasing amount of gamma energy. The light is converted to an electrical signal and processed by the sensor electronics. Scintillation detectors (refer to Figure 4-5) typically use sodium iodide crystals or a plastic material such as polyvinyl toluene (PVT) as the scintillator. While scintillation has improved detector sensitivity, this has come at a cost, their rigid design and weight.

Recent years have seen the introduction and use of flexible scintillating detectors. They differ in the material used to produce scintillation. One uses a liquid fill fluid contained within a 1″ diameter flexible cable as a scintillator. The other uses a bundle of flexible fiber plastic material. The traditional solid rod or bar scintillation detector is certainly suitable for point level detection where short lengths are used. The advantage of the flexible detector is more apparent in continuous level measurement, discussed in a later chapter.

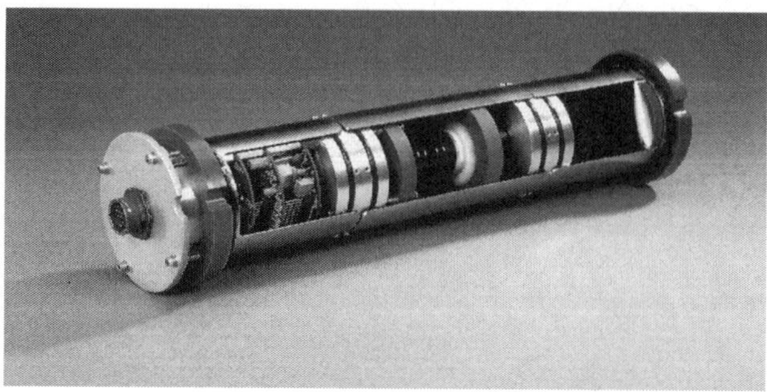

Figure 4-5   A solid rod or non-flexible scintillation detector (Ultra-Sensitive Scintillation Radiation Detector, Ronan RLL Series Brochure, Ronan Engineering Company, Measurements Division, Florence, Kentucky).

Source Material and Holder

Up to this point we have said little about the source material used within the source holder to generate the gamma radiation. There are two primary materials used, depending on the sensor, brand, and application. These are Co-60 and Ce-137. Co-60 is Cobalt-60, which has a half-life of only about 5 years. Ce-137 is Cesium-137, which has a longer half-life of 30 years. Other source materials may be used but Co-60 and Ce-137 are used most often. However, selecting the source material is not that simple. Several factors must be considered: (1) life expectancy of the source, (2) vessel wall material and thickness, (3) the target material and its density, and (4) other radiation sources local to the radiation-based level sensor that may be in use all the time or periodically. The source material type is not the only decision that needs to be made. The source size is just as critical and needs to consider the aforementioned factors in addition to the distance between the source holder and the detector, as well as the length of the detector. Source size is with regard to the amount of radioactivity or disintegrations/second that will take place. The measuring unit is milliCurie (mCi), where $1 \text{ mCi} = 3.70 \times 10^7$ disintegrations per second. The expertise for selecting and sizing the proper source material is important to the safe and proper operation of the level sensors.

Selection Considerations

You should consult with your chosen supplier and the manufacturer of the device you decide to use. A list of the pros and cons of these devices is as shown in Table 4-2.

Table 4-2 Pros and Cons of Radiation-based Point Level Sensors

Pros	Cons
Completely non-contact with process media AND non-invasive in all installations.	Much higher purchase cost than all other point level sensors ($2,000+)
No impact from materials that are extremely abrasive or corrosive.	Must maintain ambient temperature of detectors with integral electronics within specified limits, maximum 122–158°F (50–70°C); use remote electronics for high temperatures
Can be used with virtually ANY *process* temperature and pressure conditions.	Different source materials decay and have a different half-life, Cesium-137 is most common (30-year half-life).
In most situations the presence of dust during filling has minimal impact on performance of the sensor.	Potential impact of radiation to plant personnel requires planning during pre-installation and training of personnel operating in the area around radiation-based point level sensors.
Detection of material typical within 0.25″ adequate for most bulk solid applications.	Often must be licensed with Nuclear Regulatory Commission (NRC) within the United States or the Atomic Energy Commission of Canada. Ongoing testing may be required.
Internal obstructions are not a problem as they have nil impact on gamma energy passing through the vessel and material.	Material buildup and coating on inside vessel walls beyond initial calibration can provide false or inaccurate detection.
	Purchase cost, procedural, licensing, commissioning, and decommissioning costs make the total cost of ownership very high.

4.3 Conclusion

Point level monitoring or level control is a critically important application. High level detection is used to control a vessel filling process and prevent costly and dangerous overfilling. Low level monitoring and control can ensure timely notification that reordering of material is required to avoid material outages and a production shutdown. While we have discussed the use of several viable

Table 4-3   Application Features of Specialty Point Level Sensor Technologies

	Microwave Beam-Breaker	Radiation-Based Sensor
Non-invasive installation	Yes	Yes
Non-contact installation	Some	Yes
Material temperature extremes	No	Yes
Ambient temperature extremes	No	No
Abrasive materials	Yes	Yes
Large/heavy materials	Yes	Yes
Toxic/carcinogenic materials	Some	Yes
Material buildup tolerance	Non-conductive	Some
Point level control	Yes	Yes
Continuous level measurement	No	Yes
Ease of installation	Moderate	Complex
Cost of ownership	Moderate	Very high

technology choices in Chapters 2 and 3, applications where temperature, pressure, and material characteristic extremes exist necessitate non-contact or non-invasive technologies. These non-invasive or non-contact technologies do exist, but they are more expensive than the common variety of point level sensor and they involve more complicated preparation and installation. Microwave beam-breakers and radiation-based point level sensors address most, if not all, of the challenges associated with these application extremes. Refer to Table 4-3 for a summary of application features for these specialty level sensing devices.

Study Questions

Q1:   Identify and discuss the specialty point level sensor technologies covered here, as well as their primary advantage.

Q2:   Discuss the issues to consider if you are contemplating the use of radiation-based point level sensor technology.

Q3:   Are there characteristics that are common between the beam-breaker and radiation-based point level sensors?

Answers

A1:   There are specialty point level sensor technologies that should be known and understood for possible use. These are the microwave beam-breaker and the radiation-based point level sensors. The microwave beam-breaker consists of a microwave transmitter and a receiver mounted on opposite sides of the vessel and in line with each other. Because the microwave signal has a relatively wide beam in the range of approximately 26–40°, the alignment does not have to be perfect for proper operation. The transmitter emits a microwave signal, usually either of the X or K-band range, and the receiver will receive the signal without interruption or change in amplitude and frequency in the absence of material. Material presence between the transmitter and receiver will result in material detection. The primary advantage of the microwave beam-breaker is that it is a non-invasive point level sensor technology, meaning that the sensor elements do not intrude into the vessel. Being non-invasive allows the beam-breaker to be used in applications where heavy and jagged or abrasive material would damage other invasive technologies. In addition, using non-conductive windows such as glass or plastic will allow the microwave beam-breaker to be installed so that it is never in contact with the process material within the vessel, thereby increasing its resistance to abrasive, corrosive, and other problematic materials.

Radiation-based level sensing technology can be implemented as either a point level or continuous level sensor. These devices consist of a source-holder (with source material contained within), a detector, and associated electronics. The source-holder will contain a small amount of radioactive material such as Co-60 or Ce-137 that will emit gamma radiation in a controlled beam aimed at the detector. The source-holder and detector are mounted on outer opposite walls of the vessel. When properly sized, selected, and installed, the gamma energy will be emitted from the source-holder, pass through the first vessel wall, through the internal vessel environment, the opposite wall and be detected by the sensor radiation detector. The energy reduces each time it passes through either a wall or material. Therefore, after initial calibration the sensor's radiation detector will see a change in gamma energy during the presence of material within the vessel between the source-holder and detector components. The primary advantage of the radiation-based point level sensor is its complete non-contact nature. Because the source-holder and detector are mounted to the outside walls of the vessel, the vessel is never penetrated and the material and internal environment is contained. The radiation-based point level sensor is ideal for extreme problematic vessels and materials, including vessels with obstructions, process material temperature extremes, carcinogenic and other highly toxic materials, materials with high corrosion and abrasion properties.

A2:   The primary issues to consider are safety and cost. Because of these two issues, radiation-based point level sensors are often thought of as a last option scenario. As these devices utilize radioactive source material they are subject to possible licensing by either the NRC in the United States, the Atomic Energy Commission in Canada, or other similar regulating bodies in other countries. Regulatory requirements may necessitate special operational procedures for working within close proximity of these point level sensors, as well as ongoing testing to ensure safety. Licensing, special procedures, ongoing testing, and specialized personnel for installation and decommissioning are required to ensure safety of personnel and facilities.

Cost may be irrelevant if the measurement *must* be made and radiation-based technology is the only thing that will work. In fact, this is what drives part of the demand for these point level sensors. But the cost can be significant and the overall cost of ownership should not be overlooked. Purchase cost for a radiation-based point level sensor is in terms of thousands of dollars, not hundreds, typically greater than $2,000. Added to this are the licensing, installation, and commissioning costs to get the sensor up and running. Finally, any ongoing testing expenses, changes, or additions to your operating procedures to minimize personnel exposure, and the cost of decommissioning and disposal of spent source material all add up to a very large cost to own and use this technology. Critical, necessary, and extremely challenging applications is where radiation-based devices fit in the application scheme of things.

Other issues to consider when contemplating the use of a radiation-based level sensor are: (1) Does the measurement absolutely have to be made? (2) Is there any alternative technology that might perform as well or adequately enough in the application? (3) Are there any other radiation-based instruments or diagnostic tools which may be in periodic or permanent use near the proposed point level sensor location that might present interference with the level sensor?

A3:   Yes. (1) Both are non-invasive into the vessel; (2) both can be installed such that sensor components are not in contact with the material within the vessel (microwave beam-breaker requires non-conductive window into the vessel); (3) both have multiple components that require installation; (4) both are more expensive than the other common point level sensor technologies discussed in Chapters 2 and 3; (5) both must be manufactured or used within government-regulated safety guidelines, i.e. FCC and NRC; and, finally, (6) both can be problematic if not properly understood, applied, installed, and commissioned.

References

Liptak B. G., D. S. Kayser, A. J. Livingston, and J. C. Rodgers. 2003. Radiation level sensors. In *Process measurement and analysis*, ed. B. G. Liptak, Section 3.15. 4th ed. Boca Raton, FL: CRC Press.

Microwave 320 Data Sheet, Position & Level Control, PDS-D320-1B. Delavan, a division of L&J Technologies, Hillside, Illinois.

Model MWS Series, Microwave Level Switch Operation Manual. WADECO Wire Automatic Device Co., Ltd., Hyogo-Ken, Japan.

Chapter 5

Introduction to Contents Measurement

Objectives

After reviewing this chapter the reader should be able to:

- Define and discuss contents measurement for powder and bulk solid materials.
- Understand the application issues to be considered for proper choice of contents measurement strategy and technology.

Summary

Measuring the contents of a tank can be relatively simple with a liquid material, but what about a bin or silo with a powder or other granular bulk solid material? Contents measurement when the material is a bulk solid presents several challenges that do not exist in liquid applications, most of which contribute to the accuracy of the basic data that you want to know: How much material do I have? When properly taken into consideration, the attributes of the material, the vessel, and your goals can lead you to the best choice of a measurement strategy and technology.

5.1   Introduction

Measuring the quantity of material contained within your bin or silo is a function of good practices in manufacturing inventory and process control. There are as many specific types of applications as there are materials. We will not examine all of them individually. However, many fall within the category of inventory measurement and monitoring. Inventory is defined as the raw material, work-in-process material, and finished goods material within a manufacturing framework. From a business accounting perspective inventory is an asset. Each country will have its own rules regarding how to account for inventory. No matter which country, inventory is considered as an asset on the company's balance sheet. The balance sheet is where the company records the monetary value of its assets, both tangible and intangible. An asset is an economic resource because it has potential to bring value to the business by being turned into cash. Most assets are tangible, but not all. For our discussion we are dealing with tangible assets or current assets because they are deemed to be capable of being turned into cash quickly through production processing or direct sale. Inventory then is only one or two steps away from being considered cash-on-hand. Why this discussion about assets, balance sheet, and cash? Because this is the basis for understanding why the accuracy of measuring the quantity of powders and bulk solids in raw, work in progress (WIP), or finished inventory is so important. And understanding the importance in general, and specifically within your organization, has a direct impact on choosing your inventory measuring and monitoring strategy, as well as the technology you use to make the measurement. How do we go about measuring the quantity of a powder and granular material within a bin? What should our measuring strategy be? To answer that, we should understand how the bulk solid is purchased from an outside source, or produced internally.

5.2   Choosing a Measuring Strategy

Usually a bulk solid material is sold or produced by weight, but not always; sometimes it may be tendered and produced based on the volume of the material within a container. The most common measuring units for weight and volume are pounds, tons, cubic feet, and their metric counterparts kilograms, metric tons, and cubic meters. Since most bulk solids are purchased or processed based on their weight or volume, it may seem natural to measure them as they exist in inventory, in the same measuring units. This is indeed true and measuring by weight is one viable measuring strategy that we will discuss further in Chapter 7. Measuring the amount of bulk solids by weight within bins and silos of a wide variety of sizes, shapes, materials of construction, and with a myriad of material conveyance systems and other appurtenances attached to the vessels containing the material can be a viable strategy but also presents unique challenges and costs, especially since most vessels are not equipped with weighing sensors when initially purchased and installed. For this reason an alternative strategy for measuring and monitoring bulk solids inventory is to use continuous level measuring instruments to measure the amount of material based on the height of the material pile. Arguably the reason for using continuous level measuring inventory monitoring systems as an alternative strategy is that these systems often come at a much lower purchase and installed cost than weight measuring systems. In addition, the use of a continuous level measuring sensor to simply measure the empty space distance or material level may be less problematic than installing a weight measuring system on an existing vessel. Understanding the characteristics and nature of the bulk solid, its vessel, the conveyance system, and the inventory measurement goals and objectives is important to making a strategy decision, that is, to measure the bulk solid by weight or by level. In this chapter we examine these application considerations primarily from the perspective of using level sensors as the inventory measuring strategy.

5.3   Application Considerations

"Measuring" the weight of the material within a bin or silo is not always the most practical approach; sometimes it is not even viable. The weight measuring system must exclude the weight of the vessel and account for the effect anything contacting the vessel will have on its weight. But what if those bins or silos weren't installed with weighing systems? Now the vessel must be retrofitted. Not an easy or inexpensive approach. Continuous level measurement sensors and systems are a viable and cost-effective strategy (Lewis, 2010).

The Material

Characteristics of the material impact the choice of contents measurement strategy, the potential accuracy or precision of the measurement, and the selection of the most appropriate technology. Several material traits need be considered including (1) bulk density, (2) material flow properties, (3) absorbency, (4) relative permittivity or dielectric constant, (5) corrosiveness and abrasiveness, and (6) adhesion.

Bulk density was discussed in Chapter 3 in relation to its impact on the selection of point level sensor technology for a given application. Material bulk density has minimal, if any, impact on the use and accuracy of a weight measuring strategy. However, the lighter the material the more challenging the weighing of it may become using weight measuring systems. This will be discussed in Chapter 7. While the material bulk density can impact the choice of a continuous level measurement technology (to be discussed in Chapter 6), the most important impact of the material bulk density (Cocco, 2010) that needs to be understood when choosing a level measurement system for inventory or contents monitoring is with regard to the conversion of the distance/level measurement to a unit of mass or weight. Most level measuring sensors used for contents measurement will infer the material level by measuring the empty space between the sensor and the material surface at a single point on the material surface. As previously discussed, the bulk solid material is likely to be purchased, used, or produced in terms of weight, that is, pounds, tons etc.; therefore, many applications using continuous level sensors as bulk solids inventory sensors need to convert the measured distance/level to mass/weight. With the advent of microcontroller and digital electronics many continuous level sensors or systems have the ability to calculate material volume and weight based on their distance/level measurement and the vessel dimensions and material bulk density provided by the user of the vessel contents measuring system. This calculation converts from distance/level to volume and then to weight. The error in the material bulk density used for the calculation conversion directly impacts the error of the calculated weight that is reported to the manufacturing and financial management reporting system. With some materials (such as with powders) variations in bulk density within the pile of material in the vessel, and changes from the material source or production over time can affect the accuracy of the conversion calculation. The magnitude of this effect is difficult to determine. Determining what "average" bulk density to use in the calculation conversion that gives you an acceptable accuracy in the weight value, based upon empirical testing, may give you best results. In addition, periodically adjusting the bulk density value used in the calculation can also help "fine-tune" the weight calculation accuracy. If you do not know the average bulk density of the material being inventoried, consult with an inventory sensor manufacturer for their recommendation. One such recommendation (Monitor Technologies, n.d.) includes the following steps: (1) construct a sample container, (2) calculate the volume of the sample container, (3) determine compensation factor, (4) weigh bulk material sample, and (5) apply compensation factor to determine a bulk density value for the material. In addition, reference tables from third-party organizations or in trade press magazines or websites[1] may be useful.

Another impact that the material bulk density can have is on the choice of a level measurement technology. The heavier the material density, the more potential problems for level sensor technologies that measure through constant contact with the material, such as guided wave radar (TDR - time domain reflectometry) and capacitance. These level sensors have probe elements that are continuously subjected to the loading of the material on the probe based on the material density/weight. When the distance/level to be measured is great, heavy materials such as cement powder can impart high pull forces on the level sensor probes since they are suspended continuously from top to bottom in the bin or silo. The weight of the material pulling on the invasive probe elements also imparts force on the vessel roof and under some conditions may present possible damage to the sensor probe and to the vessel itself. And lightweight materials can present problems for technologies based on sound energy, as the lighter the material, the more sound energy is absorbed rather than reflected. The flow of bulk solids within vessels and conduits is a complex study in and of itself (refer to Appendix B for more information about bulk solids flow properties). While we will not cover solids flow in detail, this material characteristic can have a large impact on the accuracy of the contents measurement, therefore we must understand it well enough to properly deal with it. Proper flow of material through the filling and discharging of the vessel is essential to the overall production process.

Issues regarding material flow properties are impacted by several aspects of the material itself, along with aspects of the vessel in which the material is contained (Schulze, 2011). Material aspects affecting flow include particle shape, particle size, and the uniformity of size, chemical composition, moisture, and temperature. These aspects will determine the cohesive and adhesive forces of the material. Low cohesion between particles will generally mean that the material flows well, but stronger cohesion will lead to poor flowability and, unless dealt with effectively, will result in flow problems within the vessel. Materials that have a high adhesive nature may tend to stick to vessel walls adding to flowability problems (Figure 5-1).

Ratholing conditions occur when the material flows readily in the middle of the vessel, like a funnel, but remains in place or stagnant at the sides close to and within the cone section of the vessel. The material within the stagnant areas may not be properly measured with level measurement contents

---

[1]Bulk Density Chart. http://www.powderandbulk.com/resources/bulk_density/material_bulk_density_chart_a.htm

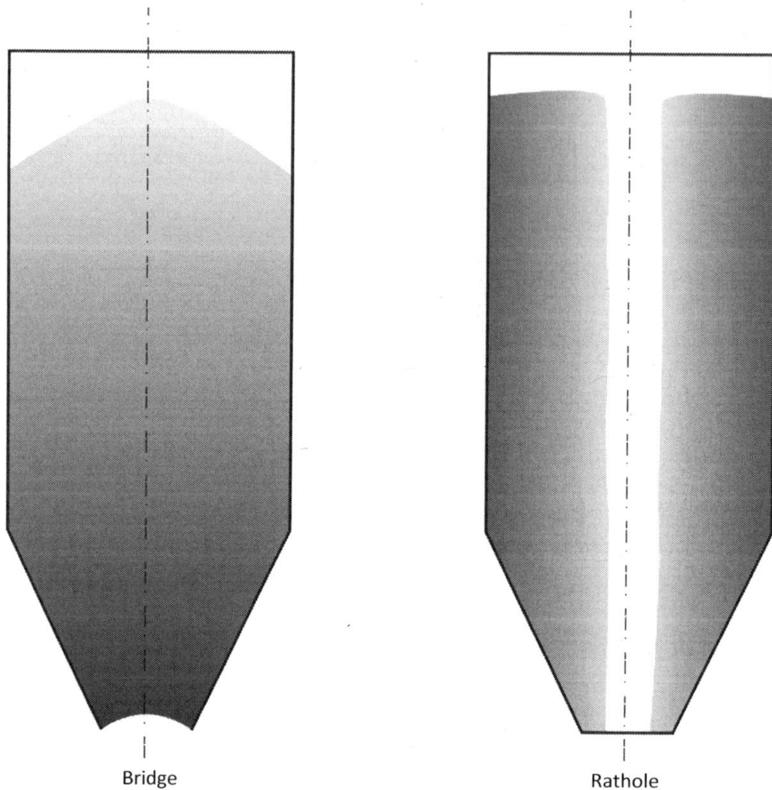

Figure 5-1   Material flow problems include ratholing and arching.

sensors, or the stagnant material can create a false measure of material at a higher level than what would more accurately represent the amount of material by weight that exists within the vessel.

Arching of material just above the discharge point is another common material flow problem. Material with high cohesive and adhesive forces, combined with vessels having too small a discharge opening is the primary cause of most arching problems. Designing vessels for good flow based on material properties and vessel needs can be done (Marinelli and Carson, 1992).

Arching and ratholing can impact inventory measuring using level measurement sensors and introduce process control problems. Traditional continuous level measuring sensors simply measure the empty space or level of the material pile at a single point on the material surface. Solving the material flow problems through whatever means available should be completed before expecting any accurate measurement of the bin contents when using level sensors. Various means exist to solve these flow problems; for example, employing aerating devices or industrial vibrators/air cannons, or through changes in the bin design itself to achieve mass flow rather than funnel flow. These flow problems will not impact inventory monitoring when using weighing systems; however, they still will affect production and process control by introducing unexpected outages and irregular flow rates.

Is the material to be monitored absorbent? Is it hygroscopic? A bulk solid material that is hygroscopic attracts and holds moisture either through absorption or adsorption. The latter is the adhesion of water molecules to the solid particle. The most common issue is absorption, which is the inclusion of the water molecule within the solid particle. Whether the material is hygroscopic, and to what extent it is, closely ties in with the flowability of the material as illustrated in Figure 5-2.

Examples of hygroscopic bulk solid materials include sugar, magnesium oxide, cement powder, fly ash, and lime. Many other examples exist. Often the flow problems that occur with these materials are due to their hygroscopic nature from a high humidity level in the environment. Measuring the contents of a bin or silo containing a hygroscopic material requires even more diligence to solve the flow problems prior to expecting an accurate inventory accounting of the material using a continuous level measurement sensor as the inventory monitor. The relative permittivity of the bulk solid material can also impact our choice of contents measurement technology. The relative

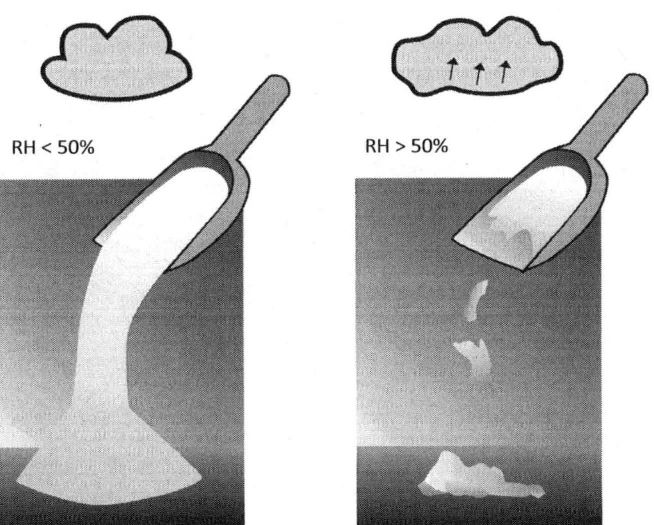

Figure 5-2 Hygroscopic material can change from a free-flowing state to being clumpy and presenting flowability problems.

permittivity of a material is also known as its dielectric constant, or ε_r mathematically (DoITPoMS, 2008). All materials can be considered non-conductive, but to what degree is what the relative permittivity or dielectric constant suggests. The measure of relative permittivity is a dimensionless number. The dielectric constant of a material will always be greater than 1.0, which is the constant for air in a vacuum. Relative permittivity, or dielectric constant, is a relative indication of how well the material will or can perform as an insulator or conductor of electricity. The relative permittivity of a material can directly impact the performance and reliability of certain continuous level contents measurement technologies. The lower the dielectric constant of the material contents of a vessel, the more challenging the measurement may be for capacitance and radar devices. The dielectric constants of a variety of materials can be found in Table 3-2. Contents with a dielectric constant of less than 2.0 (generally considered as non-conductive) can prove problematic for capacitance and radar technologies and manufacturers should be consulted and challenged regarding their device's ability to provide accurate and reliable contents measurement of these materials.

The impact the material contents have on the materials of construction of any invasive level measuring sensor technology is an aspect of the material that needs to be considered. This includes the corrosive and abrasive nature of the material. Generally speaking, corrosion has to do with the chemical compatibility between materials of construction and the contents material, especially for probe elements that will be in constant contact with the contents material. Abrasion is the process of wearing down the materials of construction of the level sensor probe in contact with the contents material by friction. How abrasive and corrosive a material is can impact the choice of sensor technology, whether contact or non-contact, and its mounting location as well. Refer to Chapter 3 for a more detailed discussion of corrosion and abrasion. Materials exhibiting high abrasion or corrosion may require externally mounted non-contact/non-invasive continuous level sensors or weighing systems as their best strategy.

Adhesion was briefly touched upon during our discussion of bulk material flow problems above. This material characteristic is the degree to which the material will tend to cling or stick to some other substance, where cohesion is the attraction between particles or molecules of the same material. Therefore, vessel contents that exhibit high adhesive properties can build up on bin walls, cone sections, and on invasive sensor elements. This buildup can create material flow problems and difficulties with either reliability or performance of a chosen level sensor. As an example, material coating the probe element of a continuous capacitance level sensor or the cable of a guided wave radar device can induce false readings where the coating is mistaken for material level, especially with materials of a high dielectric nature. Material coating the antennae of a through-air radar or acoustic sensor can also create problems for these level sensors.

The Angle of Repose

In addition to the previously discussed material characteristics, the angle of repose or shape of the material surface also impacts both choice of level sensor technology and the accuracy of calculations that convert the basic distance/level measurement to volume and then weight. In addition to material characteristics, the angle of repose of the material contents within a bin is also affected by characteristics of the bin, such as the fill and discharge locations, as well as the method of filling and discharging the material.

While measuring the level of a liquid within a vessel presents with a flat surface, making level sensor volumetric calculations easier to handle and more accurate, a bulk solid material will form a pile or create a depression (Figure 5-3). This means that the bulk material will present a sloped or angled

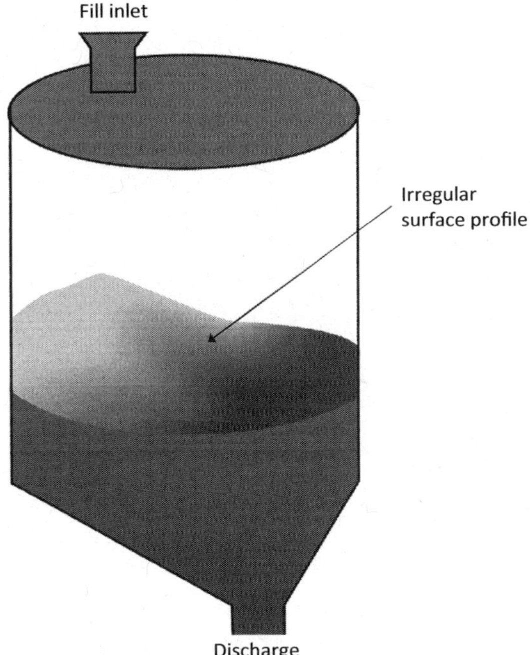

Figure 5-3   Angle of repose or material surface shape can vary widely and be irregular.

surface to the level sensor. The angled slope is commonly referred to as the angle of repose. Because level sensor technologies essentially measure the distance/level at a single point on the material surface, the angle of repose of the specific bulk solid application is of import (see Figure 5-4). Depending on where the point of measurement is on the material irregular surface, a different distance/level will be measured, therefore a different volumetric and mass calculation will be made. Ideally, accurately measuring the surface profile would allow for a more accurate volume determination (discussed further in Chapter 6).

The Vessel Containing the Material

Characteristics of the vessel containing the bulk material to be monitored can impact the accuracy of the inventory measurement when using either weight or level measurement strategies for inventory monitoring. In addition, these vessel attributes can impact the selection of sensing technology. For level

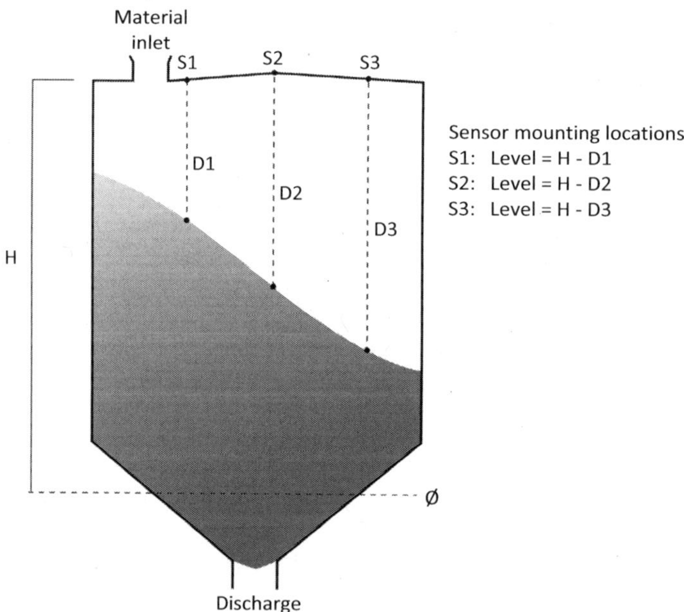

Figure 5-4   The angle of repose impacts volume calculations based on the level measurement point.

measurement sensing technologies these vessel issues include the vessel size and shape, as well as the filling and discharge of the material. Mounting, construction, and vessel isolation from other structures can impact the use and performance of weighing systems for contents inventory measurement.

A continuous level sensor measures the distance/level and then converts, sometimes using off-board software, to volume and weight. The dimensions of the vessel must be accurate if the calculation to material volume is to be accurate. Having a dimension, such as the height or diameter of a cylindrical vessel, be wrong by even 1″ can equate to hundreds or even thousands of pounds of material when the distance/level measurement is converted to volume and then weight. Vessel dimensions are best taken from the certified engineering drawings that were available when the vessel was purchased. However, checking these dimensions against the actual vessel is always a good idea. Figure 5-5 illustrates volume calculations for a cone bottom cylindrical bin and the inaccuracy of the volume calculation when the vessel diameter is wrong by +2″ and the height is incorrect by +3″.

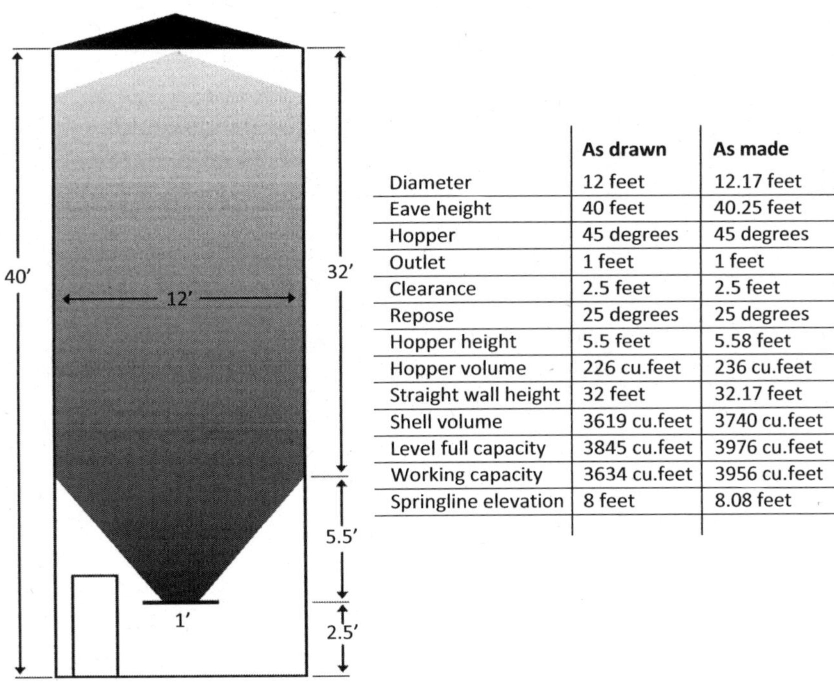

	As drawn	As made
Diameter	12 feet	12.17 feet
Eave height	40 feet	40.25 feet
Hopper	45 degrees	45 degrees
Outlet	1 feet	1 feet
Clearance	2.5 feet	2.5 feet
Repose	25 degrees	25 degrees
Hopper height	5.5 feet	5.58 feet
Hopper volume	226 cu.feet	236 cu.feet
Straight wall height	32 feet	32.17 feet
Shell volume	3619 cu.feet	3740 cu.feet
Level full capacity	3845 cu.feet	3976 cu.feet
Working capacity	3634 cu.feet	3956 cu.feet
Springline elevation	8 feet	8.08 feet

Figure 5-5   Error in vessel dimension leads to inaccurate volume calculations.

The difference between the theoretical volume of a full bin and the volume of the bin with actual "as measured" dimensions is 131 ft³.[2] If the theoretical material is polyethylene pellet at 35 lb/ft³ in density then this error would be as much as 4,585 lb! If the theoretical material were Portland cement powder at 94 lb/ft³ then the error is even more at 12,314 lb! The accuracy of the vessel dimensions is very important to the accuracy of the calculated volume and weight.

How the vessel is filled and discharged can impact the accuracy of converting the measured distance/level to volume and weight. It can also impact the choice of sensor technology and even the mounting location of the continuous level sensor. For example, if the material is pneumatically conveyed into the vessel, especially if the material is a powder, this conveying system (dilute phase conveying) introduces far greater dust into the vessel atmosphere than does gravity feeding the material into the vessel. Conveying of the material into its vessel includes a combination of material particulate and air. When the material is a powder, such as flour, cement, fly ash etc., a heavier dust concentration will exist within the empty space atmosphere, especially with pneumatic conveying. Heavy dust concentrations will impact the performance of through-air level sensor technologies such as acoustic, laser, and radar. This means that reliable measurements during filling (dynamic silo condition) may be delayed, inaccurate, or not even possible. However, inventory measurement applications may only require level sensors that can provide a reliable distance/level measurement only during static or non-filling conditions. The use of inventory sensors is never recommended for level control as well (covered in more detail later in this chapter).

The method and location of filling inlets and discharge outlets can also impact the volume calculations when using level measurement sensors for inventory monitoring of vessel contents. Filling creates a pile or a positive angle of repose of the material. Discharging the material from the vessel will create a depression or negative angle of repose. With center fill and discharge of a vessel, such as a cylindrical bin or silo, the material pile will have a somewhat regular shape in both the positive and negative direction. A "neutral" point can be easily identified for sensor distance/level measurement and its respective mounting location on the roof of the vessel, as shown in Figure 5-6, by mounting the level measurement inventory sensor so that it measures on the material surface one-sixth of the vessel diameter in from the vessel sidewall.

The "neutral" point is the point on the material surface where the volume of material above the imaginary material flat surface line is equal to the volume

---

[2]Dry Bulk Storage Capacity Calculator. http://www.powderandbulk.com/resources/capacity_calculators/dry_input.php,

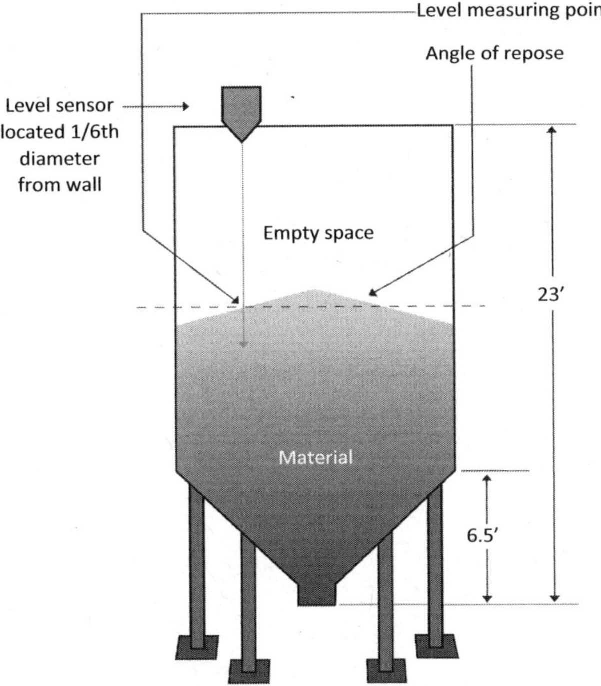

Figure 5-6 Regular shape vessels with center fill/discharge provide for easily identifiable "neutral" sensor measuring point, key to accurate volume calculation.

of the empty space below the same line. This holds true for both a positive and negative angle of repose. In addition, there is typically a transition point where the angle of repose will be relatively flat as the angle of repose changes from positive to negative. The location of fill and discharge on the vessel and the conveying details do not necessarily impact the natural angle of repose of the material, however, they do impact the ease of finding a "neutral" point where the distance/level measurement will yield the most accurate volume and weight calculations. Material cohesive, adhesive, and flow properties will impact the angle of repose, making it steep or shallow. The steeper the angle, the greater the degree of error that can be introduced into the volume calculation with irregular material surface shapes where the "neutral point" is more difficult to identify or where the "neutral" point is completely elusive. Using center fill/discharge vessels is important in bulk solids inventory applications with material that presents a very steep angle of repose, when using a level measurement inventory monitoring strategy. What can happen when vessel fill/discharge is not in the center? Imagine calculating the volume of the bin contents where the

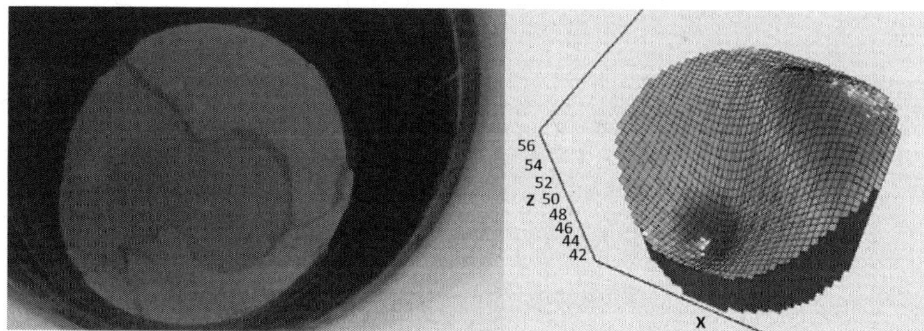

Figure 5-7   Irregular surface shape (3DLevelScanner Brochure 01-11-1M-NPC, page 3, BinMaster, Lincoln, Nebraska).

fill inlet is off-center and there are multiple discharge cones or draw-offs for the material. Figure 5-7 indicates just how irregular surface shapes can be.

Other Issues

Other issues affecting the selection or performance of continuous level sensors for inventory monitoring of powders and bulk solids include the following: the process temperature, ambient operating temperature, material corrosiveness and abrasiveness, airborne dust concentrations, and material adhesion. We have discussed these issues within this or previous chapters. However, we have not yet adequately discussed the required frequency of the inventory measurement. How often do you really need to measure your bin or vessel contents for inventory purposes? The answer to this question will help qualify what level sensor technology will best suit your needs.

Most manufacturing and inventory control systems will have a running count of parts, components, and assemblies. The inventory management system subtracts the items as they are pulled from stock and used. In this case the inventory is usually adjusted on a daily basis. If your inventory items consist of fasteners or other easily counted parts this task is not problematic. However, when the inventoried material is a powder or other bulk solid where hundreds of thousands of pounds are stored it may be a different story. Bulk solid materials are purchased, produced, or sold by volume or weight and this becomes the unit of measure desirable for inventory control and management purposes. Some bulk solid processing facilities will keep track of their inventory by weighing the material as it is used from the storage vessel. They meter the powder or granular material from the vessel discharge, or weigh it in bulk before processing. The amount metered or weighed is subtracted from the starting amount of a full vessel, perhaps 200,000 lb. In this example the only real level monitoring application is level "control," using point level sensor technology to control vessel filling and indicate a reorder as it occurs in order to prevent material outage. However, many bulk processing operations do not utilize this approach due to the additional cost of weighing and metering. Discharging the material from its storage vessel may use pneumatic conveying to transport the material from storage directly into the process without metering and weighing. Whether because of cost or material handling issues, these are the situations where level measurement sensors may be used for inventory monitoring purposes. In either case, whether metering/weighing material when discharged or using a level measurement sensor to monitor the inventory contents within a vessel, inventory records are generally updated on a daily basis, sometimes less frequently than that. Therefore, it is not far-fetched to conclude that bulk solid material inventory monitoring applications require a measurement frequency of once, twice, or perhaps as much as a few times per day at the most, and that the measurements can be made during relatively static vessel conditions (not during a filling process) since they are inventory measurements, not for control of the filling or discharge function. Applications where more frequent "continuous" inventory measurements may seem to be required are most likely those where the use of the level sensor is not strictly for inventory purposes, but for level control as well. In other words, in these situations where "continuous" second-by-second level measurements are required, not only is the objective to report periodic daily inventory measurements during usage, but also to use the inventory sensor to immediately detect and report when a filled, reorder, or outage condition exists. This hybrid application where a continuous level sensor is desired to report inventory readings and provide for level control will impact the choice of sensor technologies available for consideration in the application as the inventory sensor will need to provide a reliable measurement during both dynamic (during filling) and static vessel conditions. This can prove problematic for many level sensor technologies depending on the degree of airborne dust within the vessel during filling. However, the temptation to use continuous level sensors for this hybrid purpose should be resisted. While there are always exceptions to a rule-of-thumb, the following true story exemplifies the real danger concerning the combination of level control and inventory monitoring in a single sensor, as well as making the choice of a level sensor technology more difficult, problematic, or more expensive.

The case for *not* using the inventory sensor for vessel level control (Lewis, 2009): In the calm early morning hours on December 11, 2005, at 6:01 a.m. precisely, the fuel depot at Buncefield, England, erupted in a giant explosion made possible by the unknown overfilling of a fuel tank. Soon after the fuel had overflowed the tank it began to vaporize and at some point an ignition source combined with the growing fuel vapor cloud and the rest, as they say, is history. As with most all catastrophic industrial accidents and explosions, the Buncefield incident was meticulously investigated. More than 40 people were injured in the explosions and fire, fortunately none died. The total cost of this "accident" is estimated at close to £1 billion. The final report issued by the Buncefield Major Incident Investigation Board (2008) is an interesting and informative read. Recommendation 3 by the Major Incident Investigation Board states that a fill control system should be installed which is physically and electrically separate from the operational inventory monitoring or tank gauging system. This means keeping the inventory sensing and monitoring system (tank gauging system) separate and distinct from the fill control or level control system. To avoid critical problems, such as bulk solids vessel overfilling, it is recommended to avoid the hybrid inventory/control approach previously discussed. While a bulk solid overfill may not have the same economic impact as the Buncefield incident, it can still be very costly and dangerous depending on the material. The costs of cleanup, material loss, vessel or component damage, and possible environmental difficulties add up.

Measurement Accuracy

The real-world accuracy of the inventory measurement using a single-point continuous level measurement system for bulk solids applications is the accuracy of the level sensor, usually as a % of reading (distance or level), plus any errors associated with the calculation of volume and weight, as we previously discussed. It is not usually possible for the manufacturer of the level sensor system that is being used to monitor bulk solid inventory to tell you what the accuracy of the calculated value of volume and weight will be. Some suppliers of level sensors used for inventory measurement will estimate a typical accuracy of the calculated mass at between 5% and 10%, but no specification or warranty will typically be given. With weighing systems, the measurement accuracy offered is much better, but at a higher installed cost, as we will discuss in Chapter 7.

5.4   Conclusion

Point level monitoring or level control is a critically important application. High level detection is used to control a vessel filling process and prevent costly and dangerous overfilling. Low level monitoring and control can ensure timely notification that reordering of material is required to avoid material outages and a production shutdown. However, contents measurement has a different objective or goal, that is, to provide the periodic and accurate measure of the amount of bulk solid material held in storage and treated as an asset called inventory. The function of level control within a vessel containing a powder or bulk solid material should not necessarily be combined with inventory monitoring in a single device, especially for the filling control of the vessel. Two strategies can be employed to measure bulk solid inventory within a vessel. These are to weigh the vessel and contents or to measure the level of the material within the vessel. Each strategy has advantages and disadvantages as summarized in Table 5-1. Choosing a strategy requires careful consideration of the specific application aspects. Considering level measurement as an inventory monitoring strategy has been much reviewed in this chapter in terms of the application considerations. These considerations include both material and vessel properties that impact the choice of sensor technology. As we will see in Chapter 6, there are several viable level measurement technology choices for inventory monitoring of vessel contents, each having their

Table 5-1   Pros and Cons of Contents Measurement Strategies

Weighing Vessel Contents	Measuring Contents Level
Higher cost (refer to Chapter 7 for more detail)	Lower cost (refer to Chapter 6 for more detail)
Immune to material flow problems or other material characteristics	Technology choice requires consideration of several material characteristics such as bulk density, adhesion, flow properties, etc.
Direct measure of weight as a result of force/strain measurement	Direct measure of level by way of empty space measurement compared to vessel height from discharge to sensor mounting or measurement starting point; Indirect calculation of weight by way of indirect volume calculation and assumed density value
Installation issues on existing vessels	Relatively easy installation, depending on technology choice
Very lightweight materials may present difficulties in accurately differentiating between material contents and empty vessel weight.	Lightweight materials may be easily measured, depending on technology choice

own set of pros and cons. In conclusion, it is recommended to pursue the strategy and technology of least cost which most closely or completely meets your business requirements.

Study Questions

Q1: Identify and discuss the primary measurement strategies for bulk solid inventory measurement and monitoring.

Q2: Discuss the issues to consider if choosing level measurement as the inventory contents measurement strategy for bulk solids.

Q3: How would you decide between using a level measurement or a weight measurement solution for inventory monitoring of powder and bulk solids?

Q4: Identify and discuss other possible bulk solids inventory measurement strategies, if any.

Answers

A1: Two primary contents measurement strategies exist for inventory monitoring of powders and other bulk solids. These are level measurement and weight measurement. Ultimately, most bulk solids inventory measurement requires a value in units of weight or perhaps volume, as these are the units they are produced or purchased in. Therefore, unless directly measuring the weight of the material contents of the vessel, using level measurement sensors may require a calculation or conversion from the measured distance/level to weight. This conversion requires proper selection of the neutral measurement point to minimize the effect of the material angle of repose, and also requires accurate vessel dimensions and material bulk density value.

Level measurement offers a lower cost of contents monitoring than weight measurement; however, it may be less accurate. Weight measurement systems are more expensive and may require forethought to install them with the initial vessel purchase and during vessel construction. Weight systems provide the promise of much improved accuracy and eliminate potential problems associated with material and vessel properties that impact the performance or reliability of a level sensor system.

The use of hybrid measurement systems where the inventory sensor is also used as the sole sensor for vessel fill control is not recommended.

Due to consequences associated with overfilling of vessels it is always recommended to utilize a fill control system physically and electrically separate from the inventory monitoring sensor system.

A2: Selection of level measurement as an inventory monitoring strategy for bulk solids, and choosing the most appropriate sensor technology, requires evaluation of vessel and material characteristics including (1) accuracy of vessel geometry, (2) vessel structural and conveying issues, (3) ability to select a neutral mounting location for the sensor to eliminate or minimize the effect from the material angle of repose, (4) singular inventory sensor use or hybrid (inventory and control—not recommended), (5) material bulk density, (6) material dielectric constant or permittivity, and (7) material adhesive, abrasive, corrosive, and flow properties.

A3: There are three things to evaluate: (1) accuracy and precision goals for the measurement, (2) budget or total amount of money available for purchase, installation, and operating costs, and (3) whether the bulk solid storage vessels already exist or are to be purchased/constructed. A tradeoff can exist between accuracy/precision and installed cost. Installed cost for each strategy may vary dependent upon whether the vessel exists or is to be constructed new. Understanding the terms "accuracy" and "precision" are also important. "Precision" points to repeatability of the accuracy of the measurement, like the grouping of arrows shot at a target. "Accuracy" is simply how close to perfection the actual measurement is. Both need to be considered and achieved.

A4: "Manual" tank gauging is another alternative to level or weight measurement, and is still in use by many organizations today, though it is decreasing. This requires an employee to make a distance measurement from some point on the top of the storage vessel to a point on the surface of the material by using a measuring instrument such as a tape measure or handheld distance measuring device. This alternative is often used before purchasing and installing an automated inventory monitoring system, whether by weight or level. The employee would be required to periodically climb to the top of the vessel, remove any cover for access to the material surface, and make the measurement from the vessel top to a point on the material surface. Several issues or possible problems exist with this strategy including (1) it is labor intensive, (2) climbing to the top of the vessel may present safety concerns and have an impact on the cost of blanket liability insurance policies, (3) measurements during inclement weather may not be possible at critical production times with outdoor storage vessels or it may make these manual measurements extremely hazardous to employees, (4) additional accuracy and precision issues are introduced by the inconsistency of identifying the same measurement starting point and the point on the surface of the material where the measurement is to be made each time the measurement is done, and (5) sensing the contact of a tape on the material surface is difficult with long distances (the precision of this is also dependent upon the person performing the measurement).

References

Buncefield Major Incident Investigation Board. 2008. *The Buncefield Incident 11 December 2005: The final report of Major Incident Investigation Board.* London: Buncefield Major Incident Investigation Board.

Cocco, Ray. 2010. What is density in particle technology? *Powder and Bulk Engineering* 24 (October): 16–21.

DoITPoMS. 2008. The dielectric constant. http://www.doitpoms.ac.uk/tlplib/dielectrics/dielectric_constant.php. University of Cambridge, Department of Materials Science and Metallurgy, Teaching and Learning Packages. Revised April 2, 2009.

Lewis, Joe. 2009. Level measurement expert column: Accidents happen? *Processing* (September): 22–23.

Lewis, Joseph D. 2010. Application considerations for continuous level and inventory monitor of bulk solids. White Paper, 2nd ed. BlueLevel Technologies, Inc., Rock Falls, Illinois.

Marinelli, J., and John W. Carson. 1992. Solve solids flow problems in bins, hoppers and feeders. *Chemical Engineering Progress* 88 (5): 22–28.

Monitor Technologies LLC. n.d. Determining the bulk density of a material to improve accuracy of material weight calculation. http://www.monitortech.com/solutions/sol_bulk_density.shtml

Schulze, Dietmar. 2011. Flow properties of powders and bulk solids. www.dietmer-schulze.de/grdle1.pdf

Chapter 6

Contents Measurement: Level Technology Selection

Objectives

After reviewing this chapter the reader should be able to:

- Identify and define the basic principle of operation for the primary level measurement technologies used for contents measurement of powder and bulk solids.
- Understand the fit and use of each technology, advantages, disadvantages, and basic recommendations for each level measurement technology.

Summary

Contents measurement of bins, silos, and tanks containing bulk solids is most usually done to determine inventory levels, that is, how much of the bulk solid is in the bin. This question can be addressed by any one of three possible strategies (weight, level measurement, and manual measurement). Using a level measurement sensor to automatically measure and report the amount of material contents in a bin is one of these strategies.

Several level measurement sensor technologies exist that can be employed to measure the level of bulk solids contents within bins and silos. These technologies can be differentiated by whether they are invasive or non-invasive to the vessel, in contact or non-contact with the material, or whether the measurement is periodic or continuous (individual measurements seconds apart).

Many challenges exist in inventory measurement applications of bulk solids when using level measurement sensors. These challenges exist due to the various characteristics of the material and the vessel. Considering these challenges, as discussed previously in Chapter 5, and understanding the principle of operation of each technology and their respective pros and cons, will provide the knowledge required to identify the most acceptable level sensor technologies that can be considered for a given application. Finding the technology that can reliably measure the material contents, and meet your objectives at the lowest possible cost of ownership is the ultimate goal.

6.1　Introduction

Measuring the quantity of material contained within your bin or silo is a function of good practices in manufacturing inventory and process control, and we discussed this at length in Chapter 5. Here we will identify the level measurement sensor technologies available for measuring bulk solids inventory levels, discuss their principle of operation, and how they fit within the application maze based upon their pros and cons. Table 6-1 lists the seven commonly used level sensor technologies for bulk solid materials.

Table 6-1　Commonly Used Level Sensor Technologies for Inventory Monitoring of Bulk Solid Materials

Technology	Invasive	Contact w/Material	Periodic/ Continuous
Electromechanical	Yes	Periodic/ momentary	Periodic/ momentary
Acoustic	Yes	No	Continuous
RF capacitance	Yes	Yes	Continuous
Guided wave radar/TDR	Yes	Yes	Continuous
Through-air radar	Yes	No	Continuous
Laser	Yes	No	Continuous
Radiation-based/nuclear	No	No	Continuous

Each technology operates on a different principle and has unique characteristics, application advantages and disadvantages, as well as different price points. Most of these technologies measure the material level by inference, meaning that they directly measure the empty space within the bin and infer the level of the material based upon a user programmed total material height and the measured empty space. The exceptions to this measuring premise are the RF capacitance and radiation-based/nuclear level measurement sensors. These technologies directly measure the effect a changing material level has upon the sensor measuring technique (changing capacitance or gamma energy at the receiver).

Common to the traditional implementations of all of these level sensor technologies is that they measure the level of the bulk solid material pile (directly or indirectly) at a *single point* on the surface of the material. Devices emitting energy into the open air bin environment, such as through-air radar and acoustic devices, measure the empty space based upon the reflection of its energy, which is dispersed in a beam with some defined angle emitted from the sensor's transducer. The reflected energy may travel in an indirect reflective path back to the sensor, further affecting measurement accuracy, which is especially true with acoustic devices.

Purchase prices for the seven commonly used level sensor technologies typically range from a low of US$1000 to over US$10,000; therefore, matching your goals with your budget is important.

6.2   Level Measurement Technologies for Contents Inventory Measurement

Electromechanical

Electromechanical level sensors have been used for inventory monitoring of powders and other bulk solids for several decades. These level sensors are also known as weight and cable, plumb-bob, bin-bob, and yo-yo. These level sensors utilize a combination of electronic and mechanical technology (Lewis, 2010). One of the oldest patents for level measurement inventory monitoring of powders and bulk solids dates back 80 years, having been issued on December 21, 1926. Most notable patents in the past fifty years or so include those issued to Link-Belt Company in 1958, Rolfes Electronics Corporation in 1964, U.S. Steel in 1970, Bindicator Company in 1974, and Ludlow Industries (also know as Monitor Manufacturing, also known as Monitor Technologies) in 1979. This electromechanical technology has undergone numerous improvements and refinements.

Today these weight- and cable-based electromechanical level sensors typically incorporate both modern sophisticated electronics, including firmware designed to provide state-of-the-art automatic motor control, and an efficient mechanical system designed to maximize reliability in the application and life of the sensor. Weight- and cable-based level sensors can be set for automatic or on-demand measurement update frequency. In the automatic measurement mode they can provide the distance/level measurement update as frequently as desired, typically as frequently as every several minutes. This is more than adequate for most true inventory monitoring applications.

When a new measurement update is called for, the electromechanical weight and cable level sensor will initiate operation of its motor to lower the sensing weight/cable assembly into the bin from its mounting point on top of the bin. Upon detecting contact of the weight with the material surface the sensor electronics will reverse the motor direction and retract the sensing weight/cable system until it returns to its original position. The measured distance to the material surface is determined by the measurement of cable played out until the material surface is detected. The distance is measured in both directions. Once the distance measured is determined, sensor outputs and display (if so equipped) are updated.

The maximum measuring distance of electromechanical weight and cable sensors is typically 100 ft, but could be more depending upon the design and brand. The measuring distance is limited by the system employed to detect when the weight/cable contacts the material surface. The sensitivity required to detect the material surface increases as the distance measurement increases due to the overall weight of the sensing weight/cable being played out.

Most weight and cable electromechanical sensors use a stainless steel aircraft cable of approximately 3/64″ (1.2 mm) in diameter made of a 7 × 7 strand core (see Figure 6-1). Some brands offer a tape system where the stainless steel cable is replaced by a stainless steel tape approximately 15/32″(12 mm) wide and 0.008″ (0.2 mm) thick.

Figure 6-1   7 × 7 strand core aircraft cable 3/64″ diameter has 270 lb breaking strength.

Figure 6-2 is a picture of a typical state-of-the-art smart weight and cable inventory monitoring level sensor. Distance measurement is accomplished via either an optical pulse or hall-effect sensor counting system, depending on the brand. Modern state-of-the-art electronics and firmware allow these smart electromechanical sensors to intelligently control the sensing weight/cable travel during descent and ascent to avoid buried weight and broken cable conditions. In addition, most electromechanical weight and cable sensors today are of the self-validating type able to detect whether the measurement has been properly gathered and if a sensor failure occurred, in the unlikely event that should happen (broken cable or buried weight). The intelligent sensor technology also allows for multiple output types including serial

Figure 6-2 Typical smart weight and cable inventory monitor (the Model WC from BlueLevel Technologies, Inc., Rock Falls, Illinois).

Table 6-2   Pros and Cons of Electromechanical Weight and
Cable Level Sensors

Pros	Cons
No calibration required	Not for level control, inventory only
Excellent measuring range (up to 100 ft/30 m+)	Momentary contact of material
Low purchase cost	Moving parts, some preventive maintenance may be desirable in certain applications (heavy dust)
Simple and low-cost installation and setup	Cable or tape can be abraded by some materials
Easily maintained/repaired (if needed)	Measurement during filling is not recommended (weight could become buried in some circumstances)
Not impacted by angle of repose	Low pressure conditions ($\leq$30 psi)
Multiple outputs (communication, analog, pulse, relay-alarm)	Measured distance accuracy ($\pm$1% of reading or better) is fair

communications, such as MODBUS, analog 4–20 mA, AC or DC pulse, and relays for alarm indication. Typical measurement time is dependent on the distance measured and the sensing weight/cable system travel speed, usually from 0.7 to 2 feet per second.

Electromechanical weight and cable level sensors can also be used to measure the level of liquids and slurries. Weight and cable level sensors provide reasonable accuracy ($\pm$1% or better) and good repeatability and reliability. In bulk solid inventory measurement they can be used to measure the level of fine powders and granular materials. They are generally not affected by dust. They are limited to inventory monitoring applications where their intermittent nature and momentary material contact is acceptable. The motor used to lower and raise the sensing weight/cable system is typically not one designed for continuous duty operation. Refer to Table 6-2 for a list of the pros and cons of the electromechanical weight and cable type inventory level sensor.

Acoustic

Acoustics is the study of mechanical waves in gases, liquids, and solids, and includes sound and ultrasound. The difference between sound and ultrasound is simply the frequency of the mechanical wave. Ultrasound deals with frequencies above human hearing, typically frequencies of 20 kHz and above. In level sensor technology, sound and ultrasound fall into the same category that we identify here as acoustic. But frequency does matter, as we will discuss, so it is important to note that the acoustic sensor frequency of operation plays an important part in the application of this technology to the level measurement of a bulk solid material.

Acoustic level sensor technology is a mature technology and, although they have not been around anywhere near as long as the weight and cable level sensors, acoustic devices have been in industrial use for measuring material levels (traditionally just liquids) since the 1970s. However, even as far back as the early 1920s scientists and inventors were thinking of ways sound energy could be transmitted into a medium for the purposes of measuring flow or distance.

Acoustic level sensors are solid-state devices without mechanical moving parts other than the components used to generate the mechanical sound waves, typically piezoelectric devices. Acoustic sensors are one of several technologies using the measurement principle of time-of-flight. That is, the distance from the sensor to the material surface is directly related and calculated based upon the time it takes the acoustic energy to travel to the surface of the material, reflect, and return to the sensor transducer. This time is based upon the distance and the speed of sound through the empty space medium (usually air within vented bins and silos) in a static condition.

The operating frequency of the acoustic level sensor can vary from as low as 3 kHz to as high as 80 kHz. Figure 6-3 is a typical acoustic level sensor transducer used for powder and bulk solids level measurement. Sound energy is transmitted through the internal empty space of a bin as a longitudinal wave with a frequency, and a wavelength unique to the frequency. The frequency is how often the sound wave repeats during a defined time period such as one second. The wavelength is the distance over which each sinusoidal longitudinal wave will repeat. The higher the frequency, the shorter is the wavelength. The shorter the wavelength, the more problematic is propagation through airborne solid particles like dust because of attenuation, thus shortening (significantly in many cases) the measuring range and eliminating measurement during filling of dusty bulk solids.

In addition to challenges of measuring through dusty internal bin environments, many acoustic level sensors experience problems as a result of pressure fluctuations (this changes the speed of sound due to the compressibility of gases), changing angle of repose during filling or discharge, large particle sizes, internal vessel obstructions, as well as coating or the formation of deposits on the internal vessel surfaces. These have all presented serious

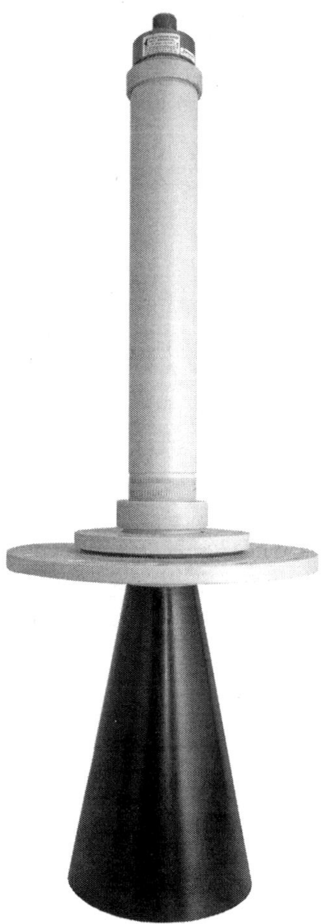

Figure 6-3   Acoustic level measurement transducer operating at 5 kHz used for bulk solids and long ranges (Courtesy Hawk Measurement Systems, Nunawading, Victoria, Australia).

challenges and problems for ultrasonic level sensing technology. Many of these issues affect the accuracy of measurement while others present disruptions in measurement due to the creation of false echoes that confuse the sensor and mask the true level measurement signal. The acoustic energy emanates from the transducer in a beam rather than a single point. The energy spreads as it travels through the empty space within the vessel. The degree of spread impacts the measurement reliability and can also impact accuracy. The larger the beam angle, the more the spread of the acoustic energy, and the more the refraction of the acoustic energy from the material surface. Multiple echoes are created and sometimes the strongest echo returning to the transducer takes a convoluted or indirect path back to the transducer with

Table 6-3   Pros and Cons of Acoustic Level Sensors

Pros	Cons
Not in contact with material	Performance affected by angle of repose
Moderate purchase cost	Can be affected by dust
Relatively long ranges	Setup requires care choosing mounting location, aiming, and parameter adjustment (on-site factory assistance may be required)
Accuracy ±0.25% of range	May require factory repair if needed
Generally rapid response to changing level	Measurement during filling may not be reliable if dusty atmosphere exists in vessel
Transducer face is considered self-cleaning	Internal bin temperature measurement needed
No moving parts	Internal vessel obstructions can impact performance

multiple reflections from other surfaces. This can impact the time-of-flight and introduce error in the measurement of the empty space. Typically the acoustic transducer will need to be aimed toward the material discharge of the vessel and be mounted well away from vessel inlets in order to optimize performance. This can present challenges when trying to get an accurate measurement at a specific point on the material surface that represents the neutral point (refer to Chapter 5) in order to improve accuracy of calculations converting distance/level to volume/weight.

Acoustic level measurement devices specifically designed for use with measuring powders and bulk solids rely heavily upon digital signal processing, much of which is unknown and patented by the manufacturers. Adjustments in programming parameters are required to optimize performance. Temperature fluctuations are not a problem. Most devices incorporate a temperature sensor and digital signal processing to compensate for the effect on the speed of sound. Table 6-3 provides a list of the pros and cons of acoustic inventory sensors.

RF Capacitance

These level sensors were developed originally for measuring liquid material levels for inventory and process level measurement applications. Of all continuous level measurement applications involving powders and other bulk solids it is estimated that RF capacitance devices represent less than 5% of the total. But whether measuring liquid level or the level of a solid the principle of operation of capacitance level sensors is the same. Figure 6-4 shows a typical RF capacitance continuous level sensor.

The RF capacitance level sensor used for inventory monitoring of a powder or other bulk solid generates a radio frequency along a sensing probe element that is invasive to the vessel and extends the distance desired for the total measuring height. The RF capacitance based level sensor operates by forming an electrical capacitor from the probe, the vessel wall, and the material contents within the vessel that is being measured. The capacitance equation is used ($C = K \cdot A/D$) where C = capacitance, K = dielectric, A = area (of the plates) and D = distance (between the two plates). The probe and the vessel wall are the two plates of the capacitor and the material being

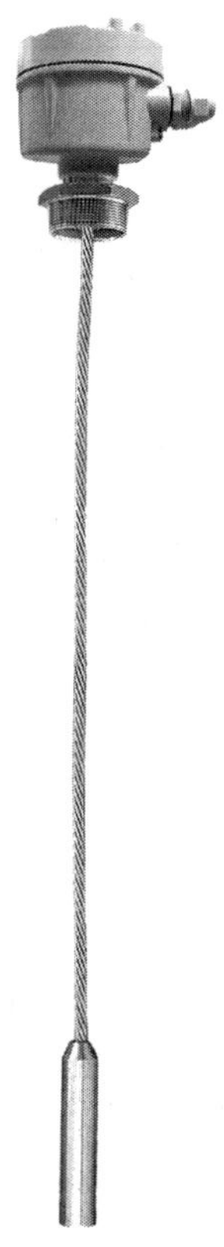

Figure 6-4   RF capacitance type continuous level sensor (Model EB RF capacitance level transmitter brochure 08-EB-B0-EP 05/15/2011, Fine-Tek Co., Ltd., Taipei, Taiwan).

measured is the dielectric. If from the equation A and D are constant, as in our applications, then the capacitance measured will vary as the dielectric varies. The dielectric will vary based on the amount of material covering the probe and the dielectric constant of the material being measured. Because the dielectric constant of the material being measured is higher than that of air in the empty space, the capacitance measured in the level sensor electronics circuit will vary as the probe element is covered or uncovered. Unlike other level sensor technologies the RF capacitance level sensor is directly measuring the effect of the level of the material pile as it increases or decreases, rather than measuring the empty space.

A pure capacitance level sensor is sensitive to material buildup or coating on the probe element (Schuler, 2001). These pure capacitance level sensors are therefore relegated to be used with clean non-coating liquids such as water. To deal with this problem and allow for further possible application in bulk solids, the RF capacitance probe is developed to measure the resistance as well as the capacitance. Typically the resistive and capacitive components of the buildup will have equal impedance, and the impedance of the buildup can be subtracted from the total impedance to obtain the impedance of only the material level. The physical length and thickness of the buildup, along with other factors, will determine the effectiveness of this procedure to ignore buildup on the capacitance level sensor probe.

One issue impacting the capacitance level sensor usage in bulk solids measuring applications is that the sensor measurement is dependent on the dielectric of the material. Because properties of bulk solids include their packing nature as they settle in the vessel, the total dielectric at the bottom of the pile can be different from that at higher points in the pile. Aeration of the material can also impact the total dielectric and measured capacitance around the level sensor probe. Discussions with the manufacturer should be thorough to be certain that the capacitance level sensor is a good choice for the bulk solids inventory monitoring application. Refer to Table 6-4 for pros and cons of RF capacitance level sensors in bulk solids inventory applications.

Table 6-4   Pros and Cons of RF Capacitance Level Sensors

Pros	Cons
Directly measures material level, not empty space	In continuous contact with material
Moderate purchase cost	Can be affected by inconsistent or changing material dielectric constant

(Continued on following page)

Table 6-4   (*Continued*)

Relatively long ranges possible (depending also on capacitance measuring range)	Installation can require mechanical attachment to bottom of vessel and field calibration (factory assistance may be required)
Generally rapid response to changing level	Measurement during filling may not be reliable if dusty
No moving parts	May require factory repair if needed
Not impacted by angle of repose	

Guided Wave Radar (TDR)

Like RF capacitance devices, guided wave radar level sensors utilize a continuously suspended cable or rod inside the vessel. This cable or rod runs the length of the entire desired measuring range. However, where RF capacitance units directly measure the level of the material, guided wave radar units measure empty space by reflected radar pulses and time-of-flight. Guided wave radar is one of four (acoustic, guided wave radar, through-air radar, and laser) level sensor technologies for inventory monitoring that use time-of-flight to directly measure the empty space and, unlike acoustic time-of-flight, with guided wave radar sensors reflecting from a specific point on the material surface is easily accomplished as that point is the intersection of the material surface and the wave guide. This makes the measurement of the empty space and material level at the "neutral" point (refer to Chapter 5) easier to accomplish with guided wave radar than acoustic.

Guided wave radar may also be known as or referred to as TDR, time domain reflectometry. TDR technology has its roots in the 1930s where it was used for geological surveying. In the 1950s TDR found promise and use being commercialized to detect breaks in telecommunication cables. However, in the early 1990s TDR was first used as a guided wave radar level sensor for measuring the level of liquids and bulk solids. Guided wave radar may also be known as MIR or micro-impulse radar.

Guided wave radar level sensors operate by generating electromagnetic energy of approximately 1 GHz, which is at the low end of what is considered by RF engineering to be microwaves. The radar energy pulses are guided toward the surface of the material by a suspended cable or rod, the length of which extends across the entire desired measuring range. The cable or rod is known as the wave guide. When the radar pulses contact the material in the vessel much of the energy is absorbed, but some is reflected. One manufacturer[1] puts it succinctly that the amount of microwave energy reflected off the material surface is related to the dielectric constant of the material. A guideline is that if the dielectric constant of the material is 2.0, then approximately 2% of the energy is reflected, the rest absorbed by the material. If the dielectric constant is 8.0, then about 8% of the energy is reflected. Here is a limitation of this level sensor technology: the lower the dielectric constant, the shorter the effective measuring range will be as a result of a lower amount of reflected energy. In spite of the small amount of reflected energy guided wave radar devices work very well and have accuracy statements as good as 0.1% of distance reading. However, given the other issues and challenges regarding overall inventory measuring accuracy (converting to volume/weight, refer to Chapter 5), this superb accuracy may be overkill.

While guided wave radar level sensors emit energy and use time-of-flight measurements of reflected energy to directly measure the empty space, there is no dispersion or refraction of the energy echo as there is with through-air technologies like acoustic or even through-air radar. This actually makes guided wave radar potentially an ideal technology for powder and bulk solids because of the angle of repose and changes in the shape of the material surface that typically exist with powders and other solids. Figure 6-5 is a typical guided wave radar level sensor used for powder and bulk solids applications. While the energy pulses emitted by guided wave radar level sensors do not disperse, spread, or have a beam angle, they do continuously exist within a defined area surrounding the wave guide probe from top to bottom and surrounding the bottom of the probe or weight at the end of a cable wave guide, or at the top in the mounting at the roof. It is critical to proper operation that the level sensor be installed so that no obstructions are present within the energy field. Any obstruction within the energy field surrounding can produce reflections resulting in false or inaccurate measurements. The energy field surrounding the wave guide varies by manufacturer and wave guide type.

Prior planning is best when considering the use of any level sensor for vessel contents measurement and guided wave radar level sensors are no exception. Critical installation requirements include choosing the level sensor mounting location, the method of mounting the sensor, and the vessel roof strength. When the manufacturers' instructions are followed, guided wave radar level sensors work well even when measuring powders during pneumatic filling (Lewis, 2007). Like most of the level measurement technologies

---

[1]Emerson Process Measurement. FAQ—Radar. www2.emersonprocess.com/EN-US/ BRANDS/ROSEMOUNT/LEVEL/ Pages/radar_faq.aspx

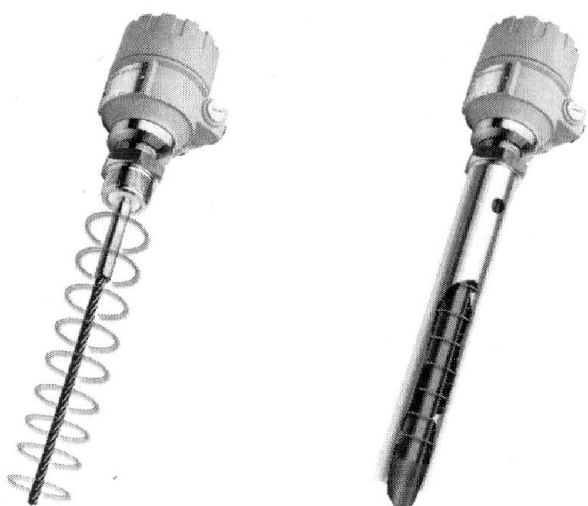

Figure 6-5 Guided wave radar for powder and bulk solids (MicroTrek brochure, front cover, Nivelco Process Control Co., Budapest, Hungary).

Table 6-5 Pros and Cons of Guided Wave Radar Level Sensors

Pros	Cons
Generally unaffected by airborne dust, angle of repose, and other process conditions	Continuously invasive and in contact with material
Very good accuracy	Material dielectric constant limits range (example: for 20 ft (6 m)+ range, dielectric constant should be >1.6 for best performance)
Responds quickly to changing material levels	Installation and material sensitive, often may require on-site factory assistance
Proven technology	Low dielectric constant material (<1.8) may require stable dielectric for best accuracy
No moving parts	Moderately expensive
	May require factory repair if needed

for powder and bulk solids there are many adjustments that can be made to the parameter settings in the electronics of these devices. Often these adjustments are critical for proper operation. In addition, installation of guided wave radar units should avoid using nozzles for mounting unless the nozzle diameter is greater than its height and the diameter is greater than the energy field surrounding the probe. Installation should also be carried out when the vessel is empty due to the probe length and the need for it to suspend the entire measuring length within the vessel. Mounting must ensure that the wave guide is far away from any possible obstruction or part of the vessel.

As with any device that utilizes a probe suspended the entire length of the desired measuring range, the wave-guide traction strength needs to match the application. The heavier the material, the more traction loading will be on the level sensor probe. The larger the diameter of the probe, the greater the amount of friction and loading will be on the probe. Traction loading[2] can exert several tons of pull force on wave guide probe elements over long ranges with heavy material. Table 6-5 provides a list of pros and cons of guided wave radar level sensors.

Through-Air Radar

RADAR (**R**adio **D**etection **A**nd **R**anging) has been with us since the mid-twentieth century, and has been used to make level measurements of various process fluids since the 1970s (frequency modulated continuous wave—FMCW) and the 1980s (pulse radar) (Devine, 2000). Radar level sensors for bin contents monitoring of bulk solids are available in two varieties: (1) pulse radar and (2) FMCW radar. Perspectives vary as to which works better for measuring the level of powders and bulk solids depending on who you ask. Pulse radar devices use the principle of the time-of-flight to directly measure the empty space between sensor and material surface, similar to acoustic, guided wave radar, and electromechanical devices that we have already discussed. FMCW based level sensors use the change in frequency of reflected radar energy to determine distance to the material surface. Both are level sensor technologies that are not brought into direct contact with the material being measured, a perceived advantage over other devices. We will take a closer look at implementations of both pulse and FMCW level sensor devices.

Pulse radar devices do not necessarily vary in appearance from their FMCW sisters (see Figure 6-6). Pulse radar level sensors use short bursts of radar energy emitted by the electronics/antennae system toward the material surface. The electromagnetic waves are typically in the C or K band, and microwave energy somewhere between 4 and 27 GHz (IEEE). As part

---

[2]Bulletin 354A Installation & Operation Flexar Guided Wave Radar Continuous Level Measurement System, Rev. 4 09182009, page 7, DOC-001-354A. Monitor Technologies LLC, Elburn, Illinois.

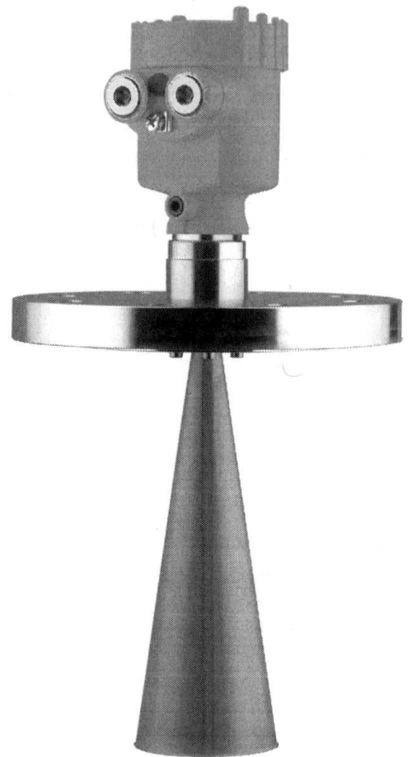

Figure 6-6   Pulse radar (through-air) level sensor (VegaPulse 68, Radar Level Measurement in Bulk Solids brochure, page 7, document 35293-US-110301, Vega Americas, Inc., Cincinnati, Ohio).

of the energy pulses are reflected off the material surface the time-of-flight of the reflections is measured and directly related to the empty space distance. An advantage of pulse radar devices, compared to FMCW, is said to be that pulse radar units directly measure time. However, measurement of time is required in nanoseconds and this presents a significant challenge. Manufacturers have devoted much time and effort to developing methods to measure very small units of time and also to attempt to factor out false echoes or reflections from internal vessel obstructions. Pulse radar can measure distance as accurate ± 2 mm (0.08″). Given the angle of repose of the surface of powder and granular materials, the need for this degree of very low measurement error is questionable.

FMCW radar was first used commercially in level measurement during the 1970s to provide high accuracy tank gauging measurement systems for the petroleum industry. Accuracy of current radar tank gauging systems

Figure 6-7 FMCW through-air radar level sensor operates at 78 GHz (Model LR560, Sitrans LR560 catalog sheet, Siemens FI 01 2009, Siemens AG, Karlshrue, Germany).

is ±0.5 mm (0.02″). Where pulse radar units measure the time-of-flight of reflected microwave pulses, FMCW radar measures the change in frequency between the transmitted frequency range and the frequency of the reflection. This frequency difference is directly related to the distance between the sensor and the material surface that reflects the microwave. Determining the delta-frequency typically requires the use of fast Fourier transformation (FFT).[3] Proponents of pulsed radar devices claim that there exist challenges to FMCW implementations, including the processing power required for the use of FFT and the challenge of ignoring false reflections from obstructions within the vessel. However, beam angles for FMCW level sensors can be as small as 3°, providing significant focus of the microwave frequency sweep and avoiding obstructions. Many FMCW level sensors operate in the K band range with frequencies between 24 and 26 GHz with accuracy of ±10 mm (0.4″) or 0.25% of reading, adequate for bin contents measurement of powder and granular material. One unit operates at 78 GHz within the W band range, allegedly for improved range and performance in heavy dust environments (Figure 6-7).

Through-air radar technology remains highly technical and specialized, requiring a tremendous investment on the part of the manufacturer to develop. It also may be difficult for many users and industry technocrats to understand the technical details of the technology, similar to acoustic or even guided wave radar. True operating range, like guided wave radar, remains dependent on the dielectric constant of the material as this impacts the amount of radar energy absorbed and reflected by the material.

---

[3]Krohne. Measuring principle: Frequency modulated continuous wave (FMCW). www.krohne.com/Measuring_Principle_Level_Radar_Level_Measurement_ en.812.0.html

Table 6-6 Pros and Cons of Through-air Radar Level Sensors

Pros	Cons
Not in contact with material	High purchase cost
High temperature applications ≤200°C (392°F) are standard	Setup may require on-site factory personnel
Fast response to changing level	Sensitive to buildup of material on sensor antennae/horn, not self-cleaning
Some reported success in long range applications with some dust	Material dielectric constant must typically be >1.5 (range dependent)
No moving parts	May require factory repair if needed

Through-air radar level sensors are sensitive to material clinging on their antennae and this situation needs to be avoided. Through-air radar level sensors are not self-cleaning and need protection or cleaning periodically if they are used in an internal vessel environment that would cause this condition to occur. Refer to Table 6-6 for a listing of the pros and cons of through-air radar level sensors for powder and granular material inventory monitoring.

Laser

The idea and development of light amplification by the stimulated emission of radiation, or LASER, first began when Albert Einstein theorized about the key process that makes lasers possible. This theory is called stimulated emission. Wikipedia defines stimulated emission as the process of transferring energy from an electron to a field of electromagnetic energy.[4] The first light laser, called a ruby laser, was invented by Theodore Maiman in 1960. Refer to Figure 6-8 for an illustration of the components of a ruby laser.

Further developments and much research continued over subsequent decades, including the development of LIDAR, **L**ight **D**etection **A**nd **R**anging, which can be used to measure distance and speed and is used for surface mapping. Today the use of laser for the measurement of the level of bulk solids is possible and commonplace for certain applications. However, the number of manufacturers is limited. It is a level sensor technology with a high barrier to entry, requiring specialized expertise and a very large long-term investment.

---

[4]www.en.wikipedia.org/wiki/Stimulated_emission

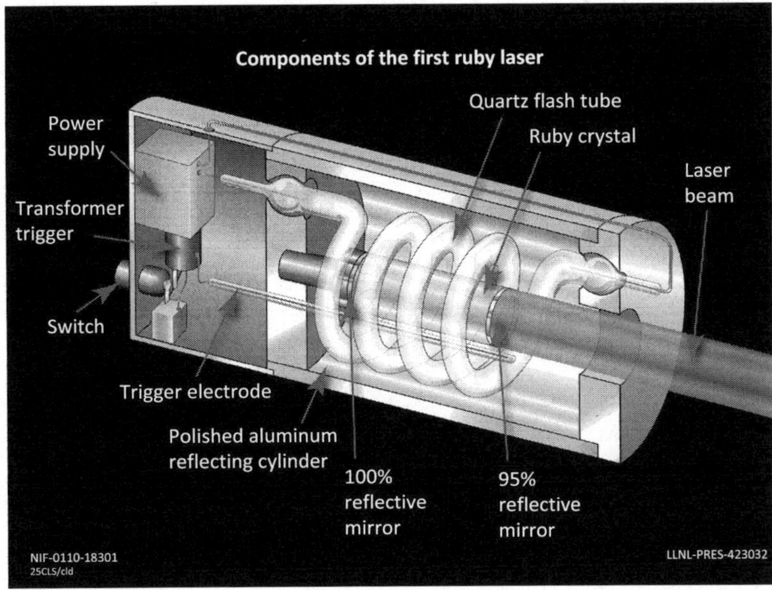

Figure 6-8   The components of a ruby laser (https://lasers.llnl.gov/education/ how_lasers_work.php, Lawrence Livermore National Laboratory, Livermore, California).

Industrial laser level sensors are non-contacting devices using infrared lasers with 905 nm wavelength. These lasers are classified regarding eye-safety using an international standard IEC 60825. There are a variety of classifications in IEC 60825. Industrial level sensors are indicated in product literature as using Class 1, 1M, or 3R lasers, depending on the manufacturer. Investigate this closely, understand the rating, and follow all manufacturers' safety recommendations and requirements.

Laser energy waves have a beam divergence. This beam divergence is similar to the specified beam angle for acoustic and radar level sensors. However, the beam divergence with laser level sensors is very small, typically less than 0.3°. This allows laser-based sensors to be used where a precise measurement point on the surface is essential, especially for avoiding obstructions. For example, consider the application of breaking up large ore or rocks for further transporting and processing. Rock breakers are used. The mined ore or rock is dumped onto a "grizzly" (see Figure 6-9), which has a screen or metal grate with openings for the broken and reduced rock and dust to fall through into the bin below. As the rock breaker reduces the size of the rock enough it will fall through the grizzly and into the bin hopper below. The broken rock and dust pile up in the bin, which acts as surge control for

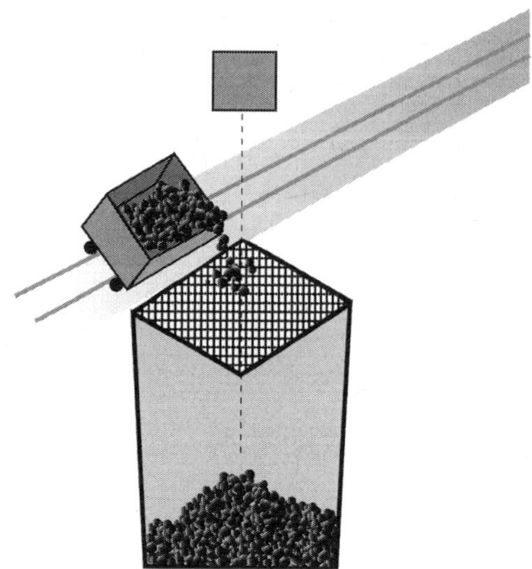

Figure 6-9   Rock breaker grizzly application (Illustration from page 1 of Product Application Note SN 0104, Optech, Inc., Vaughan, Ontario, Canada).

the transport of the broken rock to the next step in the processing of the ore. Measuring the changing level of broken rock and dust in the bin is difficult due to the grizzly grate. The level sensor must have a finely directed beam so as not to come in contact with any part of the grizzly. Non-contact is essential because of the material and the grizzly. With beam divergence of less than 0.3°, and its non-contact nature, laser seems ideal.

Laser-based material level measurement sensors operate on the principle of time-of-reflection measurement (see Figure 6-10) being directly related to the empty space measurement, similar to time-of-flight with acoustic, guided wave radar, and through-air pulse radar. Laser energy travels at the speed of light, which is a physical constant of 299,792,458 meters per second (about 186,282 miles per second). Because of this fast speed, traveling approximately one meter (39.4″) per 3 nanoseconds, the measuring time function within the laser level sensor must be very precise and with modern electronics and processing speed this is possible. Accuracy of the distance measurement can be as good as ±2 mm (0.078″).

Being a light based technology laser inventory monitoring level sensors are prone to experiencing challenges with airborne dust, the amount of which may vary by brand and model of the sensor. At least one manufacturer advertises that the empty space measurement can be made with dust in the empty space environment, specifically with "light to moderate dust." The real question is, how much is a light or moderate amount? It isn't specified or known, seemingly to be found just by trial and error. Manufacturers do

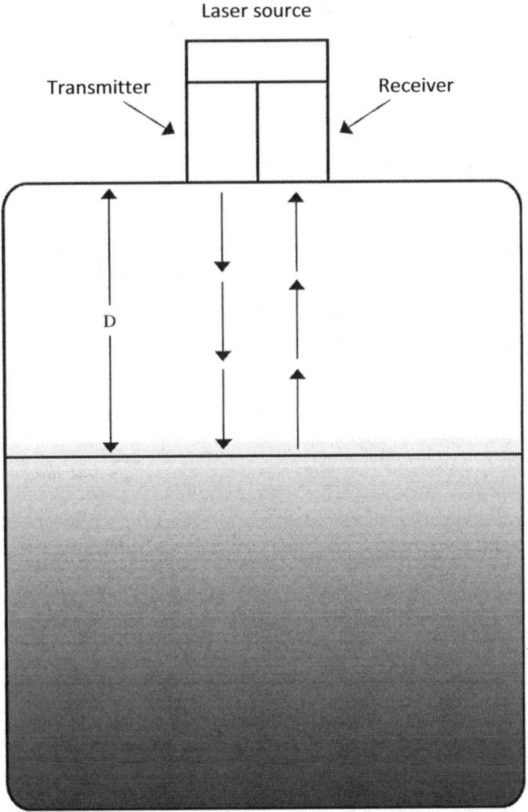

Figure 6-10   Time-of-reflection empty space distance measurement using laser level sensor.

promote and sell accessory devices such as dust tubes and air curtains to help keep dust from building up on the sensor window where it could impact performance and reliability. Digital processing techniques exist to enhance performance of some sensors to penetrate dust, the performance of which should be investigated before using this technology in moderate or heavy dust applications.

The reflectivity of the material surface being measured may also impact measurement performance or the maximum range that can be measured when using laser-based level sensors. Dark colors absorb more light than lighter ones. Shiny surfaces also may produce more reflection but the light may be reflected laterally rather than directly back to the level sensor. These reflectivity issues need to be reviewed with and considered by the manufacturer before application of the technology. Refer to Table 6-7 for the pros and cons of laser-based level measurement sensors.

Table 6-7   Pros and Cons of Laser Level Sensors

Pros	Cons
Not in contact with material, not necessarily invasive either (depends upon mounting arrangement)	High purchase cost
No real user calibration, relatively easy initial setup in applications with minimal or no airborne dust	On-site factory personnel for more challenging applications
Fast responding to changing level	Very sensitive to dust and any buildup of material on sensor lens, requires dust tubes or air curtains to help keep clean
Long range applications possible	Material color impacts range
Not impacted by angle of repose	May require factory repair if needed
Very small beam divergence, can focus measurement point	Limited operating temperature and pressures
Good accuracy	
No moving parts	

Radiation-Based/Nuclear

Radiation-based level measurement sensors have been in use for more than 30 years. They are used for both point and continuous level measurement. They are non-contacting and non-invasive sensors. As such they can be used to measure in extremely difficult environments and to measure extremely abrasive, corrosive and toxic materials. However, they are the most expensive level sensor technology available, and because of this they are generally only used where nothing else will work.

The radiation-based level sensor consists of three components, a gamma source holder, a detector, and the associated electronics for processing and transmission of the measured value (Singh, 2008). Refer to Figure 6-11 for a block diagram of a typical radiation-based level sensor. The radioactive gamma source is mounted on the outside of the vessel on one side, while the detector is mounted on the outside of the vessel on the opposite side. The installation is done so that the gamma energy is emitted from the source holder toward the detector. As the gamma energy is emitted from the source holder it passes through the vessel wall toward the detector. When no material is in the vessel between the source and the detector, the gamma

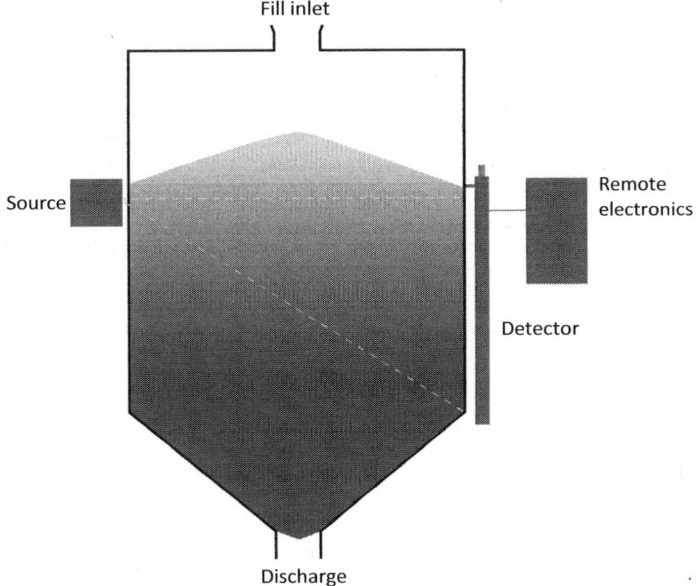

Figure 6-11 Radiation-based level sensor showing three components -source/holder, detector, and electronics.

energy passes through the opposite wall and will be detected at the detector. The walls of the vessel reduce the energy detected. The introduction of material in the vessel between the source and detector results in further reduction of the gamma energy at the detector. This reduction is compared to the reduction when the vessel is empty. This is directly related to the level of material in the vessel, as the gamma energy loss through the vessel walls is constant.

While some source holders may be capable of providing source energy for multiple vessels located directly next to each other, the measuring range for each device (source/detector/electronics) is limited to about 15–25 ft (4.5–7.6 m). Longer ranges can be measured but will require the use of multiple sensors combined to measure the entire range, further increasing the cost of measurement for the application. Measurement error for continuous level is typically ±1%.

Improvements over the years have produced radiation-based level sensors using far less radiation source material. Average working life of 1–2 decades is typical. Advances have also led to the flexible detector. Two methods are used to detect gamma energy in a flexible detector. These are a liquid scintillating fill fluid and the use of special scintillating fiber bundles.

These offer improvements in sensitivity, which has led to the reduction in the amount of gamma source or radioactive material required, as well as a sizeable reduction in the weight of the detector element from around 15 lb/ft to 1 lb/ft.

In most cases licensing and periodic testing is required for each use and site location. Some manufacturers claim to offer a level sensor system with very low gamma energy source material being required, low enough so that the periodic testing and documentation normally required by nuclear regulating bodies might be eliminated. These systems are also said to be able to be installed and removed without a licensed technician being in attendance.

Radiation-based level measurement systems have their place. An example would be the use of a radiation-based system for measuring the level in power plant fly ash hoppers. Fly ash is abrasive, reducing the life expectancy of invasive probes. In power plant applications it is common to see internal hopper temperatures of 1000°F or more, eliminating other technologies. For this application a source holder capable of providing shared source energy to multiple hoppers could be used. This may reduce overall installed costs. The use of externally mounted source and detectors eliminates maintenance issues associated with abrasion of invasive sensors and being externally mounted aids operation with very high internal hopper temperatures.

The pros and cons of radiation-based/nuclear level sensors are as shown in Table 6-8.

Table 6-8   Pros and Cons of Radiation-Based/Nuclear Level Sensors

Pros	Cons
Not in contact with material	Very high purchase cost
Not invasive into the bin	Typically has higher installation and operating costs
Useful for applications with very high operating temperatures	Licensing with nuclear regulatory agencies and specially trained personnel may be required
Useful for measuring level of highly difficult materials that cannot tolerate vessel penetration and could severely impact level sensor probes	Moderate measurement accuracy, typically ±1.0%
Not impacted by angle of repose	Relatively short range limitations

6.3   Surface Mapping Level Sensor Techniques

An inherent disadvantage of the level measurement strategy for bin contents inventory monitoring is that the measurement of empty space or material level is made at a single point on the material surface. As discussed in Chapter 5, the inventory measurement is usually desired in terms of volume or weight engineering units. Converting from a level measurement at a single point on the surface of the material has additive errors with each step from distance to weight. One large influx of error occurs during the calculation of volume from the assumption of a flat material surface based on the measurement at a single point on the material surface. These errors could be reduced or eliminated if we knew the shape of the material surface of the pile of material. In recent years technology has emerged that provides a volumetric calculation from the 3D mapping of the material surface profile for powder and bulk solids in vessels for the purpose of monitoring bin contents inventory. In addition, these systems can provide the highest financial value when using PC-based systems with either dedicated software or Internet access to a secure website providing graphical representations of the material surface and other data. Two technologies have been the basis for surface mapping devices, laser and acoustic.

Laser Devices

Laser-based level measurement scanning and mapping sensor systems utilize laser distance measuring sensors and a multiple-axis rotating mechanical mounting system to obtain distance measurements from multiple points on the surface of the material. These types of systems are relatively new, with US Patent 6,986,294 dating to a filing in August 2002, with issue date of January 2006. These laser-based surface mapping systems use a Class 2 laser that is considered to be eye-safe because the blink reflex will limit exposure if directly looking at the laser light.[5] However, there is a warning message regarding staring into the laser light. The multiple-axis system moves or rotates the laser position and aim in two directions. The two-axis movement allows the scanner to make multiple laser-based distance measurements and create a three-dimensional image. Limitations of this system includes those inherent with laser level sensor technology, including the sensitivity to measuring during filling with dust; therefore, mechanical means to prevent dust from building up on laser lens is needed (sensor includes a door that closes

---

[5]BinTech Precise Inventory Measurement brochure. www.bin-tech.com/Index_Files/ Docs/3D%20Inventory%20Scanner%20Brochure.pdf

when the sensor is not in operation, thereby protecting the lens from any dust and dirt), and measuring during static non-filling conditions is best. Accuracy claims of inventory measurements compared to data from weigh scales are 0.5% error.[6] The laser surface scanning device is applicable for use in very large bins with long ranges where errors converting to volume from using a level measuring sensor that measures at a single point on the material surface can be very high, or where multiple single-point level sensors are needed and accuracy of the volume of material remains questionable as well. Information from these scanning systems is typically available on a desktop computer with visual illustrations of the material surface within the bin.

Acoustic Devices

Acoustic-based surface mapping or scanning sensor systems use multiple acoustic transducers to measure distances in multiple directions (see Figure 6–12). Significant proprietary (International Patent WO/2006/090394

Figure 6-12   Acoustic-based 3D scanner which maps material surface (3D LevelScanner brochure, front cover, BinMaster, Lincoln, Nebraska).

---

[6]http://www.bin-tech.com/index_files/Page310.htm

filed February 2006, US Patent Application 20090007627 filed February 27, 2006) approaches are used by the manufacturer, including the signal processing that produces surface mapping and volumetric data viewable using specific PC-based software. Transducer frequency is said to be very low, less than 4 kHz, which is claimed to negate the problems traditional acoustic or ultrasonic level sensors have with dust during filling (Perl, 2010). Using a very large 70° beam angle a significant number of echoes in three dimensions are received every 5 seconds. The time of each acoustic pulse and its resultant echo is known and captured for processing. The time-of-flight method of distance measurement is applied to each pulse/echo and related to the distance of the pulse/echo as well. Refer to Table 6-9 for a list of the Pros and Cons of Surface Mapping level measurement technologies.

Table 6-9   Pros and Cons of Surface Mapping or Surface Scanning Level Sensors

Pros	Cons
Not in contact with material	Very high purchase cost
Provides calculated bin contents measurement values	Higher installation and operating costs where factory-based support may be needed, possibly on-site or via remote linking
Good accuracy of volume, perhaps between ±0.5% and 1.5%, eliminates most errors associated with calculating volume based on single point distance or level measurement	Laser-based system is sensitive to dusting and any buildup of material on sensor lens requires covering and use during static conditions; acoustic-based system may be more tolerant of dust and use during filling
Useful for measuring material contents of very large bins, where multiple single-point level sensors would be needed	Factory repair required if needed
Not impacted by angle of repose	Limited operating temperature and pressures
Surface map image produced can also alert and show material flow and buildup problems within the bin	

6.4   Conclusion

Continuous distance/level measurement for bin contents inventory monitoring purposes does not necessarily require measuring in real time during filling (refer to Chapter 5). Therefore, monitoring bin inventory contents may be able to utilize technologies that can suffer performance or range reductions or reliability issues due to internal dust during filling, assuming they are applicable for measurement of the material within the bin during static conditions. Inherently technologies that do not come in direct continuous contact with the material surface may be preferred, and for some applications this non-contact feature may be required to prevent perceived or real contamination of the material. However, in reality this may only be necessary in a small percentage of applications and may come with higher costs as non-contact technologies are higher in purchase cost than the periodic contact electromechanical sensors, and even the non-contact version of radar technology implementation is more expensive than its contact counterpart, guided wave radar or TDR. As stated in Chapter 5, it is recommended to pursue the technology of least cost that most closely or completely meets your business requirements. Table 6-10 is a guide to aid in identifying possible technologies for use in bin contents inventory monitoring through the use of level measurement sensors.

Eliminating the errors associated with conversion from distance/level to volume and weight may be important in some applications, as this error (refer to Chapter 5) can be significant. In these applications the use of surface mapping 3D distance scanning devices may be appropriate. However, they come with a higher price tag than traditional single point continuous level sensors, some more expensive than even the radiation-based technology previously discussed. However, with dramatic improvements in inventory accuracy in volumetric terms this expense may be justified.

Study Questions

Q1:   Identify and discuss the primary level measurement technologies for bulk solid inventory measurement and monitoring.

Q2:   Discuss the issues to consider when selecting a level measurement sensor for inventory contents measurement of bulk solids.

Q3:   What is meant by single-point level measurement sensors? Why is this important and what alternatives, if any, are there?

Table 6-10   Guide and comparison of key attributes for single-point measurement inventory monitoring

Technology	Non-contact	Non-invasive	Sensitive to Dust	Sensitive to Angle of Repose	Dielectric Sensitivity	Price
Electromechanical	Periodic and momentary	No	No	No	No	Low
Acoustic	Yes	No	Yes	Yes	No	Moderate
RF capacitance	No	No	Yes	No	Yes	Moderate
Guided wave radar	No	No	No	No	Yes	Moderate
Through-air radar	Yes	No	Some	Some	Yes	High
Laser	Yes	Some	Yes	No	No (but color)	High
Radiation-based	Yes	Yes	No	No	No	Very high

Answers

A1:   There are seven primary level measurement sensor technologies to choose from for powder and bulk solid bin contents inventory monitoring. These include weight and cable electromechanical sensors, acoustic sensors, RF capacitance, guided wave radar (TDR), through-air radar (pulsed and FMCW), laser, and radiation-based nuclear level sensors. All of these technologies measure distance or material level directly, based on a single-point location on the material surface. Each has its own set of pros and cons and the choice of the best technology depends upon the specific application. For only inventory measurement purposes where periodic measurement of the material during static conditions (non-filling) is needed, generally any of these technologies may be suitable depending on preference, internal bin temperatures, and budget.

Electromechanical units operate by measuring the distance to the material surface as a weight and cable sensing system is lowered and raised. Measurement of distance is done by counting pulses from either optic or hall-effect sensing systems that represent a specific cable distance. Once a measurement cycle is completed the sensor outputs (serial communication, pulse and analog) are updated with the latest measurement and the sensor remains idle until another measurement is called for manually or by its automated programmed measuring sequence. These units are fairly accurate and cost-effective.

Slightly more expensive is the acoustic-based level sensor. This sensor uses technology that generates sound pulses of a variety of frequencies. The sound pulses are directed toward the material surface. Echoes are produced in numerous directions when the sound contacts the material surface. Echoes find their way back to the acoustic transducer and through digital signal processing the correct echo from the material surface is selected and the time-of-flight is measured and related to a distance. These level sensors are not in contact with the material surface, but are invasive into the vessel. They do tend to be self-cleaning but inaccuracies of the distance measurement can result from indirect reflection paths of the echoes. In addition, because the sound energy spreads out with a specific beam angle the sound reflects off other objects in the bin and these echoes can be confusing or indicate a false level measurement. Adjustments and proper aiming of the transducer toward the vessel outlet can be important. Accuracy appears adequate but could be questionable due to the reflected path of echoes.

RF capacitance level sensors are similar in price to the acoustic devices, depending upon the distance to be measured. They are in continuous contact with the material and the probe element suspends within the vessel for the entire measuring range. Because of cost and weight they may be limited in practical range. The RF level sensors advantages include that they directly measure material level, unlike many of the other devices that infer level based upon their direct distance measurement and a known vessel height. They operate by detecting the effect of the material dielectric on the sensor probe. They are sensitive to the material dielectric constant and generally require installed calibration for best accuracy. They can be problematic in applications with dust or where material clings and builds up on the level sensor probe.

Guided wave radar level sensors use the principles of TDR (time domain reflectometry) to measure the time-of-flight of radar frequency pulses guided to the material surface by a cable or rod probe suspended the entire length of the desired measuring range. These units directly measure distance and infer level. They are sensitive to material dielectric constant, which impacts range and measurability of the material. They tend to be more expensive than electromechanical devices and slightly more than acoustic level sensors depending upon range and brand. They do not tend to be sensitive to dust within the vessel empty space or to normal buildup on the sensor probe unless it is very heavy and conductive.

Through-air radar technology can be either pulsed radar or FMCW (frequency modulated continuous wave). Pulsed radar units measure time-of-flight of emitted radar pulses and their reflections back to the sensor transducer. The energy is emitted with a very small beam angle and this means that the angle of repose has slight or very little impact on the measurement. The sensor is not in contact with the material surface but is invasive, and buildup of the material from dust on the sensor antennae can impact performance and reliability. Like other solid-state sensors they may be prone to requiring setup and installation assistance from factory-based or specially trained personnel. Through-air radar sensors have good accuracy but are high priced, above guided wave radar, acoustic, RF, and electromechanical technologies.

Laser and radiation-based nuclear level sensors are also available. These two technologies are less used and more expensive, radiation-based being the most expensive. However, both do not contact the material surface and may also be installed so that they are not invasive to the bin. Laser level sensors measure time-of-flight reflection of the laser light energy. Like radar these reflections are generally measured in nanoseconds so sophisticated electronics and digital signal processing is required, which adds to the expense of the sensors. Radiation-based sensors use radioactive gamma wave source material in a source holder, a radiation detector, and electronics to measure the material level based on changing gamma energy detected. They are very expensive and installation can require regulatory agency licensing, ongoing testing and inspections, and installation and setup by specially qualified and licensed personnel.

A2:   Selection of the most appropriate strategy for a given inventory or bin contents measuring application requires consideration of vessel and material characteristics (refer to Chapter 5) including: (1) accuracy of vessel geometry, (2) vessel structural and conveying issues, (3) ability to select a neutral mounting location for the sensor to eliminate or minimize the effect from the material angle of repose, (4) singular inventory sensor use or hybrid (inventory and control, which is not recommended), (5) material bulk density, (6) material dielectric constant or permittivity, and (7) material adhesive, abrasive, corrosive, and flow properties. In addition to material and vessel properties just mentioned, selection of the most appropriate level measuring sensor should generally consider the following:

(a) Is a non-contact sensor required? If it is "desired" but not required, what additional cost might it be worth in the application if any? Is a periodic momentary or short duration contact with the material acceptable?

(b) Is it acceptable for the sensor probe element to be invasive to the vessel (sticking into the vessel)?

(c) Is measuring during filling *required*? Note that typical inventory-only applications need to measure only during static conditions unless material level changes very rapidly.

(d) Does the angle of repose change or will its specific angle impact the technology from making a reasonably accurate distance or level measurement?

(e) Does the material within the vessel have a dielectric constant that may rule out using those technologies that are dependent upon material dielectric properties, such as RF capacitance and radar?

(f) Are there any other aspects of the material that may eliminate either of the level sensor technologies, such as material buildup properties and magnitude, abrasion, corrosiveness?

(g) What is your price sensitivity or budget? Generally speaking, find the least expensive technology that will meet your needs for accuracy, reliability, and installation and overall ownership costs.

A3: The seven level measurement sensor technologies reviewed essentially measure the empty space or material level at a single point on the surface of the material. For the purposes of converting this distance/level measurement to volume and weight, a flat surface is assumed. It is important to make the measurement at a neutral point (refer to Chapter 5) or errors introduced into the conversion calculation could be significant. Depending upon the location and number of fill inlets and discharge outlets, the surface shape may be very irregular. This makes the location of a neutral point very difficult. An alternative or solution for errors introduced by assumption of a flat surface is to be able to measure the surface profile or shape and calculate volume and weight based upon this profile. Surface scanning level sensors exist that do just this. However, the major tradeoff is expense as these sensors and systems can be very costly. However, for applications requiring very high accuracy and precision they may be the best choice and a great alternative to weight measuring systems.

References

Devine, Peter. 2000. History of radar. In *Radar level measurement: The user's guide*, Chapter 1. www.vega.com/downloads/Radar_book_chapter1.pdf. West Sussex, England: VEGA Controls Ltd.

Lewis, Joseph D. 2007. Ensuring successful use of guided-wave radar level measurement technology. *Powder Bulk Solids* (April).

Lewis, Joseph D. 2010. Application considerations for continuous level and inventory monitoring of powders and bulk solids. White Paper, 2nd ed., BlueLevel Technologies, Inc., Rock Falls, Illinois.

Perl, Ofir (CEO, APM Automation Solutions Ltd., Tel Aviv, Israel). 2010. Volume measurement using 3D technology. Presented at the POWID conference, Mumbai, India, May 2010.

Schuler, Edward. 2001. *A practical guide to radio frequency level controls*. Horsham, PA: AMETEK Drexelbrook.

Singh, S. K. 2008. *Industrial instrumentation and control*. 3rd ed. Uttar Pradesh, India: Tata McGraw-Hill Publishing Company Ltd.

Chapter 7

Contents Measurement: Weighing Technology

Objectives

After reviewing this chapter the reader should be able to:

- Identify the primary categories of weighing technologies used to measure bulk solid material contents in bins and silos.
- Understand the operation, advantages, and disadvantages of each category of weighing technology.

Summary

Weighing the contents of powder and granular material in storage bins and silos is a direct measurement in the units the material was most likely produced or purchased in. For this reason, it is a desirable method of contents measurement. It is one of three strategies that can be employed to monitor the inventory of bulk solid contents within these vessels. However, the difficulty and expense involved with retrofitting existing vessels makes this strategy questionable in many situations. The installation and use of level measurement sensors appears easier, less costly, and straightforward. However, level measurement will require conversion from a distance/level measurement to volume and weight. This conversion can include errors introduced through incorrect dimensions and assumptions (refer to Chapter 5). Where high accuracy of the contents measurement is absolutely essential, the use of weighing technology can be advantageous, and perhaps required no matter what the expense.

7.1   Introduction

Measuring the quantity of material contained within your bin or silo is a function of good practices in manufacturing inventory and process control, and we discussed this at length in Chapter 5. This measurement may be desired in terms of weight or mass, such as pounds or tons. One strategy for bin contents measurement is the use of weighing technology. Two categories of weighing technology exist: (1) stationary load sensor systems and (2) pressed-in/ bolted-on strain sensor systems designed for retrofits. In this chapter we will identify the principle of operation of these systems, their advantages, and disadvantages. It should be noted once again that the use of inventory monitoring sensor technology should not replace the use of level control devices, especially high level control sensors to be used to shutdown the filling process to prevent overfill.

The history of weighing dates back several thousand years with beginnings before the Egyptian civilization (Sanders, 1960). In addition, the first weighing systems changed little for thousands of years, other than refinements and minor improvements. The Egyptians refined weight scales and then little else was done to effectively improve them for thousands of years. The measure of distance began by using human body parts for standards of measure. For example, a cubit was the distance from a man's elbow to the tip of his fingers, and the fathom was the span from the outstretched fingertips on one hand to the outstretched fingertips on the other hand. It was not until the 1500s that the modern day metric system has its roots. In 1795 France adopted a system of measure that included the meter for a measure of length, liter for liquid volume, and gram for a unit of mass. This system was further refined, evolved, and adopted over the centuries.

While distance measurement found its early system based on human body parts, the measure of weight could not. In simple terms the measure of weight is a simple name given to the effect of force on an object as a result of the effect of gravity. This is why the weight of a body changes when the gravitational effect changes; the weight of a body on earth of 100 lb will be 16.55 lb on the earth's moon. In scientific terms the words "weight" and "mass" are not the same. However, often in our everyday use they are used interchangeably. Where

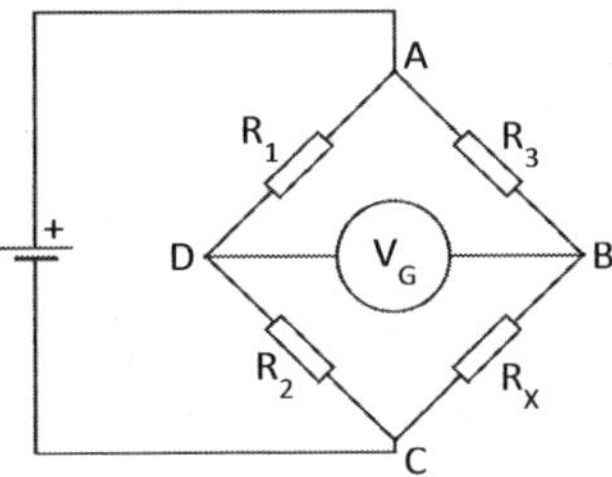

Figure 7-1   Wheatstone bridge circuit.

"weight" is the force on a body as a result of gravity, "mass" is an inherent property of matter (matter is anything that has mass and occupies volume). For our purposes we may use the terms interchangeably; however, weight is what we are discussing. As the weigh scale began evolving many centuries B.C., so did weights. Most early weight standards came from nature, the use of seeds, grains etc. The seed of the carob tree was originally used for measuring the weight of gemstones and gold, including diamonds, and this is the source for the term we know today as the carat, a weight of measure for diamonds.

In 1833 the forerunner to the Wheatstone bridge was developed by Samuel Christie, a British scientist and mathematician. He developed what he called the "diamond method" that measured the resistance values in metal wires with varying thicknesses and dimensions. Later Sir Charles Wheatstone proposed and advanced Christie's methods in 1843. The Wheatstone bridge is an electric circuit used to measure an unknown resistance, R_x, as shown in Figure 7-1. The input is usually several volts and the output measured as V_g is in millivolts. Today the "load cell" is in common use for weight measurement. The most commonly used type of load or force cell is the strain gauge.[1] The bonded wire strain gauge was invented in 1938. In 1952 the bonded foil strain gauge was invented and this type of sensor remains a leading type of strain sensor for load cell use. A strain gauge is a sensor whose resistance varies as force is applied to an object that the strain gauge is attached to, resulting in deformation of the object. The deformation can be compression, elongation, etc. The combination of a Wheatstone bridge type circuit for measuring resistance, and the bonded foil strain gauge to produce a variable resistance due to the load on a mechanical structure by the weight of bulk material in a bin or silo is the essential concept of measuring the weight of

---

[1]Introduction to strain gages. Omega Engineering Technical Reference, http://www.omega.com/prodinfo/straingages.html

the material. In addition to foil strain gauge devices, the first semiconductor strain gauge device was introduced in the 1970s and has evolved since then. Two styles of modern day weight measuring elements are in use today. While manufacturers may use different names we will generically refer to these categories as the stationary type weight measuring system that is typically mounted beneath the legs of the vessel, and the pressed-in/bolted-on strain gauge sensor based weight measuring system.

7.2   Stationary Weight Measuring Sensor Systems

The stationary weight measuring sensor is a commonly used weighing sensor for dry bulk materials in hoppers, bins, and silos of many sizes. Many weight measuring load sensors are based on semiconductor strain gauge technology, but not necessarily all (see Figure 7-2).

Load sensors are available with varying capacities. The larger the weight sensing capacity, typically the larger is the physical structure of the load

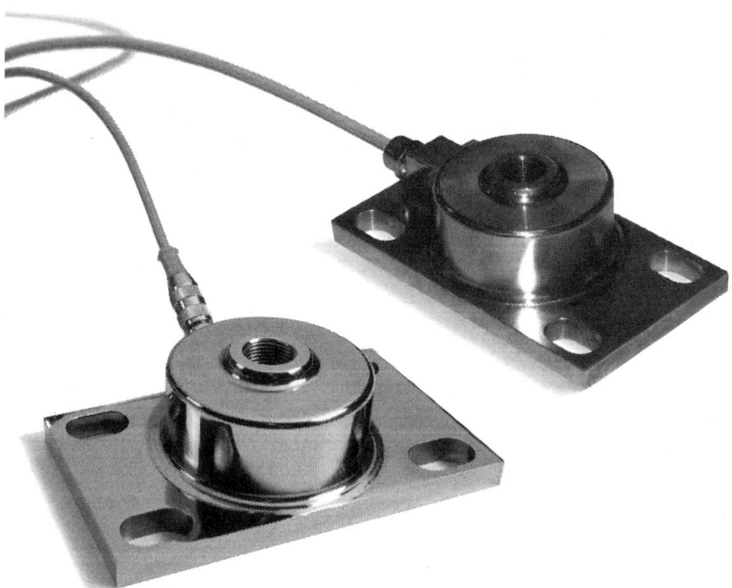

Figure 7-2   LD3/360 weight measuring load sensor using semiconductor strain gauge technology (Load Disc LD3/LD360s Installation Manual 97-1137-01 Rev. E, February 2006, front cover, Kistler-Morse, Venture Development Corporation, Spartanburg, South Carolina).

sensor. The sensor shown in Figure 7-2 is available with a maximum capacity of 25,000 lb per sensor. Figure 7-3 is a weight measuring stationary load sensor with capacities up to 1,000,000 lb per sensor. Stationary types of weight sensors are installed underneath the legs of the bin, silo, or hopper and therefore measure the load due to compression.

There are different methods of attaching the load sensor underneath each support leg, such as the illustration shown in Figure 7-4, and this depends also upon the size and configuration of the stationary weight sensor packaging. In the illustration in Figure 7-4 the use of an adapter plate attached to the top of the weight measuring load sensor is shown. Each leg of the bin is then bolted to the adapter plate as the bin is lowered onto the load sensors after they have been secured in place. Installation of the load sensors requires careful mechanical construction or retrofitting. Proper leveling is required.

The load sensors will output an electrical signal proportional to increasing or decreasing material contents in the bin. The electrical signal is in mV DC

Figure 7-3 Load Stand II in use, large capacity stationary weight measuring load sensor (Technical Specifications Load Stand II, PN 97-7006-01 Rev. D, Kistler-Morse, Venture Development Corporation, Spartanburg, South Carolina).

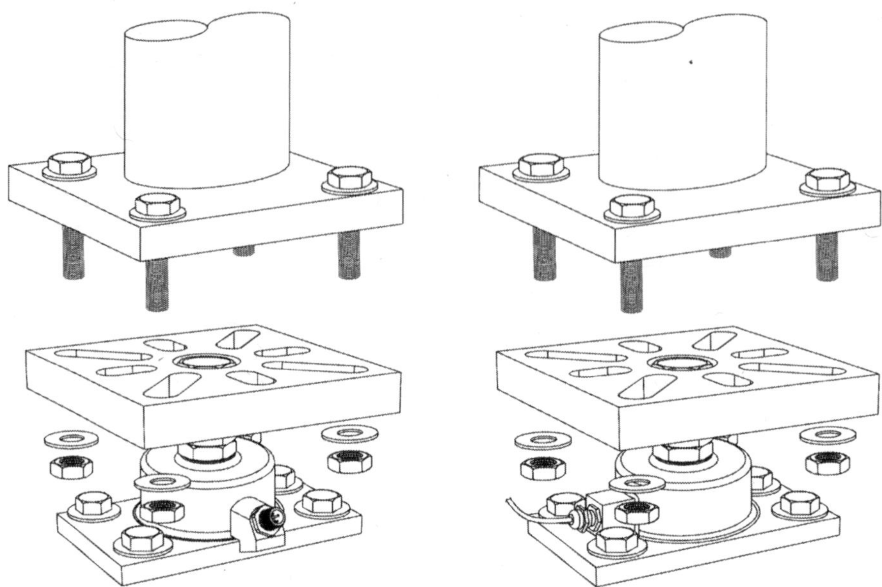

Figure 7-4 Load sensors with adapter plate and bin lowering onto sensor assemblies (Load Disc LD3/LD360s Installation Manual, Figure 3-4, page 3–4, Kistler-Morse, Venture Development Corporation, Spartanburg, South Carolina).

and the electrical signal for all load sensors supporting the vessel is tied into a weight display/controller device directly or through one or multiple junction boxes. Calibration and setup of the weight display/controller is very important for accuracy and reliability. This process requires planning, knowledge, and perhaps even prior experience. For this reason many weigh system suppliers offer setup and installation services. Indication of weight information and processing of sensor signals, along with some setup and calibration is typically done using a display indicator/controller device. These are available in generic form or from the manufacturer or supplier of the weight measuring load sensors. Compatibility between sensors and display indicator/controller is important. Figure 7-5 shows a display indicator/controller. These display indicator/controllers may also be used for the same function with the pressed-in/bolted-on weight measuring strain gauges (section later in this chapter). Basic functions of a weighing system display indicator/controller include (1) to provide voltage to weight measuring load sensors, (2) to provide operator interface to read weight and to make normal adjustments and setup of instrument and weighing system, (3) to perform functions for calibrating the weighing system, and (4) to control output interfaces including serial, analog, and

Figure 7-5 SVS 2000 weight display indicator/controller instrument interfaces with load sensors, provides operator interface and signal processing (Technical Specifications, PN 97-7052 Rev. C, front cover, Kistler-Morse, Venture Development Corporation, Spartanburg, South Carolina).

discrete outputs of weight and other data. Before use of the weighing system can begin the weight measuring load sensors need to be tested and the system needs to be calibrated.[2] Instructions are typically provided by the manufacturer of the sensor and display indicator/controller. A variety of methods may exist for system calibration. Given the cost of purchasing the hardware, installation, and setup it is advisable to calibrate the system for the highest possible accuracy. This usually requires setting a zero point with the vessel empty and then adding a known quantity of as high an amount as possible.

A stationary weight measuring load system can provide very high accuracy. The amount of error depends upon many things, including the quality and performance of the sensors and display indicator/controller, installation, and the system calibration. It is possible to have a total system error of only 0.5% of weight. However, this approach to bin contents inventory monitoring is not for the faint of heart when it comes to cost. A typical weighing system using stationary weight measuring load sensors can cost upwards of $50,000 depending on vessel capacity, desired accuracy, installation methods, calibration, and whether

---

[2]SVS 2000® installation and operation manual 97-1129-01 Rev. B May 2000. Kistler-Morse, Venture Development Corporation, Spartanburg, South Carolina.

Table 7-1 Pros and Cons of Stationary Weighing Systems

Pros	Cons
Non-invasive and non-contact with material, good for sanitary applications	Very high purchase and installation cost, especially in retrofit applications
Works with material flow problems	Installation and calibration expense is high and often requires factory assistance
Not affected by dust or material buildup	Difficult and expensive to retrofit vessels
Safe for handling hazardous materials	Potential accuracy problems if material load is less than weight of the vessel
Higher accuracy of weight measurement versus converting a level sensor measurement to volume and then weight	Structures connected to vessel can affect overall performance as can anything that impacts load of vessel
Wide measuring range from 100 to 1,000,000 pounds	Complex system of components; performance impacted by connection to adjacent vessels or structures of any kind

a new or retrofit installation is required. Stationary compression type load cell weighing systems are used when you must have the highest possible accuracy of measurement in terms of weight, and you can justify the expense. They need to be considered prior to installation of the bin or silo and then be installed with the vessel. Retrofitting an existing vessel with a compression type stationary weighing system is difficult and costly. Table 7-1 provides a general list of the pros and cons of the stationary weight measuring sensor systems.

7.3 Pressed-In/Bolted-On Strain Gauge Type Sensor System

These weight measuring systems utilize strain sensors that are attached to the frame or support members of the vessel being weighed, rather than using load cells mounted underneath the vessel legs or supporting structure as with the stationary systems. The primary reason for using the pressed-in/bolted-on strain gauge weighing sensor system is the relative ease of retrofitting an existing bin or silo. The sensors used for these weighing systems attach to vessel support members. As weight is added to or taken out of the vessel, the members that make up the supporting structure of the vessel incur deformation as a result. Strain is a measure of the deformation. The name "strain gauge" is appropriate for the device that measures the amount of strain. Stress, sometimes confused with strain, is the measure of the internal forces at work within the deformed mechanical member, not a measure of the amount of deformation itself. With that brief definition of stress and strain we leave the subject of continuum mechanics behind.

Some strain sensor weighing systems do actually "bolt" on to the vessel members, while some attach by being "pressed-in" or inserted into a drilled hole in the structure members (see Figure 7-6). Both types (pressed-in and bolted-on) claim advantages over the other, specifically in the area of temperature compensation, which is an important issue. The quantity of sensors and their specific mounting locations are dependent upon vessel type and support structure. An accurate survey of each specific vessel to be weighed is important. This information will be required by the weighing system supplier in order to recommend sensors and placement. As with the stationary weighing

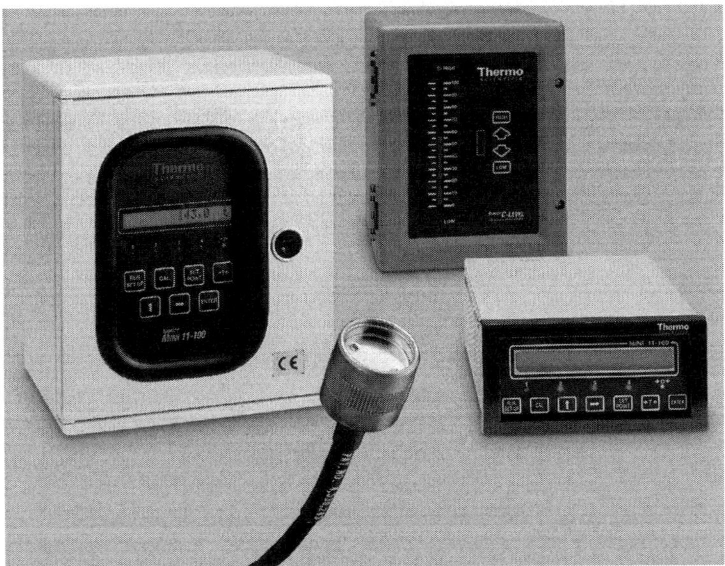

Figure 7-6 Pressed-in strain gauge sensor with controls, used in strain gauge type weighing system attached to support structure (C-LEVEL stainless steel strain sensor with controls, product specifications, front cover, Thermo Fisher Scientific, Inc., Minneapolis, Minnesota).

system, the pressed-in/bolted-on strain gauge weighing system consists of multiple sensors, wiring, junction boxes, and a display/controller device. The sensors require testing after installation and the entire system must be calibrated before use in order to deliver promised or reasonable accuracy.

These weighing systems tend to be cost-effective ($2,500 and up plus installation and calibration expenses), by comparison to stationary weight measuring systems, but still more expensive overall than many level measuring instrument technologies. They address the need for cost-effectively retrofitting existing bins and silos with a direct weight measuring system. System measurement accuracy is typically 2–5% of weight, not necessarily as good as the stationary system and perhaps a little better than the best installations using level measurement sensors. The pros and cons of the pressed-in/bolted-on weighing system are shown in Table 7-2.

Table 7-2   Pros and Cons of Pressed-In/Bolted-On Strain Gauge Weighing Systems

Pros	Cons
Non-invasive and non-contact, all external to vessel so good for sanitary and difficult materials	High purchase and installation cost, but lower than stationary weighing systems
Easier to install than stationary load cell weight measuring systems	Location of strain sensors and calibration is critical to performance and can be difficult; requires factory trained personnel for best performance
Works in applications where material flow is a problem	System accuracy typically is 2–5% by weight
Not affected by internal vessel environment	Accuracy limitations if material load is very small in comparison to the vessel weight
Safe for handling hazardous materials	Other structures attached to vessel can produce accuracy problems
Cost-effective when high accuracy of weight measurement is not required	
Ideal for retrofitting existing vessels where weight measurement is required	

Carefully examine the real-world improvement in accuracy of a strain sensor weighing system versus the calculated weight based on a level measurement system. The accuracy improvement of the strain gauge pressed-in/bolted-on system may not justify the expense of installation and calibration. However, for vessels where any invasive sensor cannot function properly or where flow problems exist and cannot be resolved, these weighing systems offer a viable solution, especially for retrofitting existing vessels.

7.4   Conclusion

Weight measuring technology is a viable strategy for inventory measurement of solids in bins, hoppers, and silos. Forethought and planning is necessary for achieving the best return on investment in weighing systems. Thinking about bin contents measurement before you purchase a bin or silo is always best as retrofitting is not always ideal or possible. We reviewed the two types of weighing systems: (1) stationary type weight measuring systems and (2) pressed-in/bolted-on strain gauge type weight measuring systems.

Table 7-3   Comparing Level Sensor and Weighing Systems

	Level Sensor Systems	Stationary Weighing Systems	Pressed-In/ Bolted-On Strain Gauge Weighing Systems
Ease of retrofitting existing vessels	Simple	Difficult	Moderate
Bin contents inventory measurement affected by material flow problems	Yes	No	No
Sensitive to material characteristics	Some	No	No
Suitability for sanitary applications	Some	Yes	Yes
Accuracy (by weight)	Low–moderate	High	Moderate
Installation complexity	Low	Moderate–high	Moderate–high
Installed cost	Low–moderate	Very high	Moderate–high

Stationary weighing systems are best for new installations where the vessel has not yet been installed or for existing systems where the highest degree of weight measurement accuracy is required, no matter what the cost. The pressed-in/bolted-on strain measuring weighing systems are easier to install on existing vessels and less expensive than the stationary system. This makes them ideal as an alternative to level measurement sensors. They are usually still more expensive to purchase and install than a level sensor, but they provide direct weight measurement with accuracy by weight better than most scenarios using level sensors. Table 7-3 provides a general comparison of key attributes and issues of weight measuring and level sensor systems.

Study Questions

Q1:   Identify and discuss the weight measuring systems available for use in bin contents inventory measurement of bulk solids.

Q2:   Discuss the issues to consider when choosing between a level measurement sensor system and a weight measuring system for inventory measurement of bulk solids.

Q3:   What are the salient features of stationary and pressed-in/bolted-on weight measuring systems?

Answers

A1:   Weight measuring systems for inventory measurement of bulk solids in bins and silos fall into two categories: (1) stationary type load cell based weighing systems installed under the vessel mounting and (2) pressed-in/ bolted-on weighing systems attached to vessel supports. Both use strain sensors, and both are systems made of multiple components including multiple sensors, wiring, junction boxes, and a display/controller device. In addition, both types of weighing systems have the advantage over level sensor based inventory measurement because they do not ever come in contact with the internal bin contents and are not affected by material characteristics or flow problems. Weighing systems also have an inherent advantage in that they are measuring in terms of weight, which is typically the desired unit of measure for bulk solids inventory, while a level sensor based system measures empty space or material level and then requires a calculated conversion from level to volume and then to weight. Therefore, weighing systems eliminate the error associated with the calculation including assumption of a flat surface on the material pile.

Stationary systems are the most accurate, as good as 0.5%, and the most expensive. They are best applied to new bins being installed rather than the retrofitting of existing vessels as this becomes very expensive. Pressed-in/bolted-on systems are designed for easier retrofitting of vessels and can also be installed on new vessels where the high accuracy of a stationary system is not required or the higher cost of a stationary system is to be avoided. The installed cost of pressed-in/bolted-on systems are more expensive compared to a level sensor based inventory system. However, their accuracy is better than the typical level sensor system at between 2% and 5% by weight.

A2: There are several issues to consider when contemplating the strategy for inventory measurement of a bulk solid material in bins and silos: (1) accuracy of measurement required/needed, (2) budget available, (3) new vessels or retrofit, (4) complexity of installation and setup, (5) support personnel required, and (6) whether a weighing system can be installed on the bin.

What accuracy is required and truly needed for your inventory measurement? Is this measurement purely for accounting purposes or are there other objectives? Level measurement sensor suppliers should be able to commit to the accuracy of their supplied sensor, but accuracy of any calculated volume and weight is something outside their purview and control. Under the best of circumstances the accuracy by weight of a level sensor based system could range from 2% to 20%; many suppliers would state that the calculated weight is an estimate or perhaps typically within 10% of actual. If measurement accuracy by weight is needed and if the required accuracy is far better than this, then a weighing system should be considered.

But weighing systems are not inexpensive to purchase and install. A realistic budget needs to be developed. If the expense of the hardware, installation, and calibration is not affordable, then consider a level measuring system that has a less expensive installed cost and compromise on the measurement accuracy. Whichever you choose, make sure you have a reasonable and accurate budget dedicated to the task.

Consider whether you are installing the inventory bin contents measuring system on a new vessel, or whether you need to install the measuring system on an existing vessel. This, due to cost, may help focus your decision especially if ±2% accuracy is acceptable. This would allow lesser expensive pressed-in/bolted-on weighing systems or level measurement sensor systems to be used.

Make sure you have the personnel qualified and capable of installing and setting up your chosen system, weighing or level. Weight measurement requires a system of components and a rather complex calibration and setup. Level measurement systems often can be handled by your personnel or perhaps with a quick visit from the manufacturer. Weighing systems being more complex can require longer timeframes to install and calibrate.

The final point to consider is whether or not your vessel can be weighed. Large grain bins with flat bottoms typically cannot be weighed. Therefore, level sensor based inventory monitoring systems are usually considered for these bins.

A3:   Stationary and pressed-in/bolted-on weight measuring systems share some features and advantages over level measurement sensor based systems, including:

(a) They are not affected by material flow problems,

(b) They are not affected by material characteristics and are typically suitable for difficult materials, whether corrosive, abrasive, toxic, carcinogenic, or any other material issue,

(c) They are very suitable for sanitary applications, including food and pharmaceutical, as they never come in contact with the internal vessel contents or environment,

(d) They provide a measure of the resultant strain imposed by the weight of the material in the vessel as it changes rather than just an empty space distance or level of the material pile that must be converted to weight with inaccurate assumptions, etc.

Stationary and pressed-in/bolted-on weighing systems also have differences, including: (1) pressed-in/bolted-on systems are more easily installed, especially when retrofitting existing vessels; stationary systems are best used on new vessel installations; (2) stationary systems are capable of providing the highest degree of accuracy, perhaps as good as 0.5%, while pressed-in/bolted-on systems provide good system accuracy of between 2% and 5%; and (3) the purchase, installation, and setup costs are typically very different.

Reference

Sanders, L. 1960. *A short history of weighing*. Birmingham, England: W. & T. Avery Limited. (First pub. 1947.)

Chapter 8

Conclusions

The subject of level measurement and monitoring of bulk solids is important for proper processing, production, and inventory control. It affects a bulk handling or processing company in a wide variety of ways. Employing the most appropriate technology for level control and inventory monitoring can provide many benefits, including the following:

1. Control of vessel filling to ensure optimal material level at a full condition.
2. Ensure filling shutoff to avoid overfill conditions and the resultant problems.
3. Ensure production activities at optimal conditions by avoiding material outage conditions.
4. Improve production efficiencies by detecting plugged chute conditions when they occur.
5. Optimize material replenishment.
6. Provide a consistent and repeatable method of inventory accounting of bulk solids in storage vessels.

Achieving these and other benefits requires an understanding of the application characteristics and technologies available for level sensing. Bulk solid materials have a substantially different behavior within a vessel than do liquids due to the nature and characteristic of both liquid and solid states. We have identified and discussed the unique characteristics of bulk solid materials as they exist within their containment vessels. Uneven material surface, material flow characteristics and problems, and the seeming compressibility of bulk solids within their vessel must be properly understood and considered. These unique application characteristics demand our attention if we are to select the best level measurement sensor to meet the objectives of each specific application.

8.1   Level Control

The use of point level sensor technologies is typically for the purpose of level control, such as the point level sensor shown in Figure 8-1 controlling the powder level in a coating machine for electric motor armatures.

There are seven technologies used for detecting the presence and absence of bulk solid materials. These are tilt, pressure sensitive diaphragm, rotary paddle, capacitive proximity, RF admittance/capacitance, vibrating element, beam-breaker, and radiometric point level sensors. Understanding the principle of operation, advantages, and disadvantages of each technology, as discussed in Chapter 2, assists in making the best choice. The application issues to review when considering technology choices include: (1) process temperature, (2) mounting location, (3) bulk density, (4) particle size, (5) dielectric constant, (6) corrosion and abrasion characteristics, and (7) adhesion properties. Table 8-1 shown below, and shown and discussed in Chapter 3, provides a summary guide for technology selection based on typical application characteristics.

Figure 8-1   Vibrating element level sensor controls powder coat material level to within a consistent height of 2 mm for repeatable high quality coating, *BlueLevel Technologies, Inc., Rock Falls, Illinois.*

Table 8-1 Technology Selection Guide Based on Common Application Characteristics

	Diaphragm	Tilt	Rotary	RF	Vibrating	Prox
Material						
Powder	Y	Y	Y	Y	Y	Y
Granular	Y	Y	Y	Y	Y	Y
Slurry	N	N	CF	Y	CF	CF
Liquid	N	N	N	Y	CF	Y
Bulk Density						
Very low 1.5–5 lb/ft^3 (24–80 kg/m^3)	N	N	CF	CF	Y	CF
Low >5 lb/ft^3 (>80 kg/m^3)	CF	CF	Y	CF	Y	CF
Medium >15 lb/ft^3 (>240.3 kg/m^3)	Y	Y	Y	Y	Y	Y
High >40 lb/ft^3 (>640.7 kg/m^3)	Y	Y	Y	Y	CF	Y

(Continued on following page)

Table 8-1   *(Continued)*

Moisture						
Low	Y	Y	Y	Y	Y	Y
High	Y	Y	Y	Y	Y	Y
Process Temperature						
>200°F (93°C)	Y	Y	Y	CF	CF	N
>300°F (150°C)	CF	N	CF	CF	CF	N
Material Coating						
Minimal	Y	Y	Y	Y	CF	CF
Heavy	N	CF	CF	Y	N	N

Y = Yes; N = No; CF = Consult factory.

8.2   Bin Contents Measurement

Measuring the contents of a tank can be relatively simple with a liquid material, but what about a bin or silo with a powder or other granular bulk solid material? Contents measurement when the material is a bulk solid presents several challenges that do not exist in liquid applications, most of which contribute to the accuracy of the basic data that you require: How much material do I have? When properly taken into consideration, the attributes of the material, the vessel, and your goals can lead you to the best choice of a measurement strategy and technology.

The best approach to measuring the bulk solid contents of a bin or silo for inventory management purposes is going to be the least expensive method that best meets your needs. There are three strategies; manual measurement, measurement using level measuring sensors, and bin contents measurement using direct weight measuring systems. These strategies provide for wide-ranging reliability, repeatability, and accuracy and have a varying range of installed and operating expense. Choose wisely; the most expensive is not always the best choice.

Measuring bin contents manually requires personnel to climb to the top of the bin or silo, open a man hatch or other entryway into the vessel on the roof and then make a measurement of the empty space from the top to the material surface. In addition to the obvious labor cost and safety issues, this method is unlikely to be very repeatable and can be problematic to determine contact with the material surface, especially when the empty space is very great. Eliminating the repeating labor costs, safety issues, measurement repeatability, and accuracy problems is easy. Use either level measuring sensors or weight measurement systems.

Level measurement sensors for use in monitoring bulk solids inventories provide reasonably accurate measures of material levels and empty space. These measurements are very repeatable. There are several technologies available, each with specific advantages and disadvantages. A summary of these technologies and how they may fit within the variety of applications in industry is shown in Table 8-2. The available technologies include electromechanical (weight- and cable-based sensors), acoustic (ultrasonic, acoustic, to etc.), RF admittance or capacitance, radar (both guided wave and through-air types), laser, and radiation-based devices.

The primary difficulty or challenge using level measurement sensors is with regard to accurately converting the material level measurement to weight, when this is required. The level measurement sensors do not have any idea what the material surface shape looks like. Therefore conversion calculations will assume a flat surface (invalid assumption with potential for errors) and

Table 8-2   Guide and Comparison of Key Attributes for Single-Point Measurement Inventory Monitoring

Technology	Non-contact	Non-invasive	Sensitive to Dust	Sensitive to Angle of Repose	Dielectric Sensitivity	Price
Electromechanical	Periodic	No	No	No	No	Low
Acoustic	Yes	No	Yes	Yes	No	Moderate
RF capacitance/ admittance	No	No	Yes	No	Yes	Moderate
Guided wave radar	No	No	No	No	Yes	Moderate
Through-air radar	Yes	No	Some	Some	Yes	High
Laser	Yes	Some	Yes	No	No (but color)	High
Radiation-based	Yes	Yes	No	No	No	Very High

calculate the volume the material occupies based upon this assumption as well as based upon the dimensional data of the vessel. Dimensional information is also potentially in error. The next step is converting the volume calculation to weight by applying a bulk density factor. What is the bulk density? Answering that question is also a challenge with potential errors due to varying densities of material during seasonal changes or different loads. Density within the pile may differ between the top of the pile of material and the bottom due to the seeming compressibility of the material. Weight calculations based on the level measurement may vary widely in terms of their accuracy and precision. However, level measurement sensors automate the process of inventory monitoring and offer very good accuracy of the material level and empty space. They are also very repeatable and typically less expensive in comparison to the alternative, that is, weight measurement. The difference in measurement performance between the level sensor technologies does not vary widely. Preferences, expense, and contact versus non-contact requirements are the primary factors in decision making, outside of application issues.

Two types of weight measuring systems exist for use with bin contents inventory measurement. These are the stationary weight system using load cells in permanently installed locations underneath the bin legs. The other type of weight measuring system is the pressed-in/bolted-on strain gauge system that measures non-compression load or strain on the bin structure members. Both types of weight measuring systems consist of the load or strain sensor that converts force into an electrical signal, wiring from the multiple sensors, junction boxes, and the remote display/controller that makes sense of the electrical signals and provides a readout display for the operator in weight terms. The pressed-in/bolted-on system is best used for retrofitting existing vessels already in place and in operation. Load compression type systems provide the best measurement accuracy and are recommended primarily where the highest possible accuracy is required and the vessel is new and not yet installed. Weighing systems require greater installation expense budgets due to their installation and calibration sophistication. Both measuring the contents material level and measuring the contents weight directly are automated approaches. The sensor systems automatically make measurements at a varying frequency and provide electrical signals or display updates for operator or control system usage. Table 8-3, also shown in Chapter 5, indicates the pros and cons of each strategy.

8.3   Conclusion

So how do we wrap this up? Level measurement is one of the so-called five horsemen of process measurement, that is, level, pressure, temperature,

Table 8-3   Pros and Cons of Contents Measurement Strategies

Weighing Vessel Contents	Measuring Contents Level
Higher cost (refer to Chapter 7 for more detail)	Lower cost (refer to Chapter 6 for more detail)
Immune to material flow problems or other material characteristics	Technology choice requires consideration of several material characteristics such as bulk density, adhesion, flow properties, etc.
Direct measurement of weight as result of force/strain	Direct measure of level by way of empty space measurement compared to vessel height from discharge to sensor mounting or measurement starting point; Indirect calculation of weight by way of indirect volume calculation
Installation issues on existing vessels	Relatively easy installation, depending on technology choice
Very lightweight materials may present difficulties accurately differentiating between material contents and empty vessel weight.	Lightweight materials may be easily measured, depending on technology choice

flow, and composition measurements. While there are nearly four times the number of liquid level measurement applications than there are powder and granular material applications, the level measurement and monitoring of powders and other bulk solids provides very unique challenges not found in liquid level measurement applications. Most of these challenges are related to material characteristics and flow properties. However, that does not mean that level measurement and monitoring applications involving powder and granular materials are more difficult to assess and address with measurement instrumentation. In fact, whether for control or inventory measurement purposes it is relatively easy to find your way through the application and technology minefield. The problem you may find is the absence of a single reference handbook to provide concise guidance. In the previous chapters we have walked through the minefields and summarized what you need to know in order to address bulk solids level measurement applications. Finding the best technology for addressing your goals and needs with the least possible installed expense is usually the best choice. Please review the chapters and appendices as needed and feel free to contact the author with comments and questions at any time.

Appendix A
Safety

1 Introduction

This appendix provides information regarding two topics related to plant and facility safety. The goal of this appendix is to provide this information as an educational tool and to promote awareness of plant safety. The information contained herein in no way replaces local, state, and federal safety laws, requirements, and codes that take precedent at all times.

The two topics covered within this Appendix are reprinted from white papers titled "An Introduction to Safety Instrumented Systems and Safety Integrity Levels" and "An Introduction to Hazardous Electrical Locations" with permission from BlueLevel Technologies, Inc., Rock Falls, Illinois.

2 An Introduction to Safety Instrumented Systems and Safety Integrity Levels

Background

It was a calm, crisp day the morning of December 11, 2005, in southeast England. Being Sunday morning most citizens were off from their work and enjoying the remaining hour of their sleep, or perhaps beginning their ablutions in preparation to go to their local Church meeting. At 6:01 a.m. local time in Hertfordshire, England many lives were changed forever when the first explosion rocked the Buncefield oil storage depot. It was said that at least one of the initial explosions was of "massive proportion" the likes of which England had not seen since World War II.

At the Buncefield depot area significant damage occurred to both commercial and residential property in the area, and 20 fuel storage tanks and the majority of the site were engulfed in a massive fire that burned for several days. Forty-three people were injured, 2,000 people were evacuated from the immediate surrounding area, and it is estimated that the cost impact was nearly £1 billion ($1.6 billion). By God's grace there were no fatalities.

What happened? The Major Incident Investigation Board (MIIB) that investigated the Buncefield accident published its final report at the end of 2008. Volume 1 of the report contains the detail, this being over 100 pages in total length. Late in the afternoon of Saturday, December 10, the day before the explosion, a fuel delivery began to arrive at one of the tanks at the Buncefield facility. Unfortunately, the safety systems to shut off the supply of the fuel to prevent overfilling of the tank failed to operate. As overfilling continued, about 10% of the 300 tons of fuel that had overfilled the tank turned to vapor. The vapor concentration grew and became capable of supporting combustion. The vapor cloud was visible on security film and witnesses near the facility reported seeing it. A severe explosion resulted and a massive fire followed.

Safety Instrumented System

The investigation authorities for the Buncefield accident also indicated that appropriate safety level systems should be employed to prevent this type of accident from occurring again and they referred to IEC 61511 as the standard involved with these types of safety systems.

Actually there are two standards issued in regards to process safety, IEC 61508 and IEC 61511. IEC 61508 preceded 61511 and is targeted primarily at suppliers or manufacturers of equipment. IEC 61508 specifies Safety Integrity Levels (SIL) for systems and devices within Safety Instrumented Systems (SIS) based on the probability of a failure of the device.

To meet a given SIL, the device needs to have less than a specified probability of dangerous failures according to the standard, and also have greater than the specified safe failures. Failure probabilities are calculated by performing a Failure Modes and Effects Analysis (FMEA).

Probability of Failure on Demand (PFD) and Risk Reduction Factor (RRF) for different SILs as defined in IEC 61508 are as follows in Table A-1.

The SIL requirements define what needs to be done to prevent systematic failures from being introduced into the device or system during design. These requirements can be met by establishing a very rigorous development process

Table A-1   PFD and RRF Based on Safety Integrity Level

SIL	PFD	RRF
1	0.1–0.01	10–100
2	0.01–0.001	100–1,000
3	0.001–0.0001	1,000–10,000
4	0.0001–0.00001	10,000–100,000

or by establishing that the device has a sufficient history of operation in order to argue that it has been proven by field use.

Electric and electronic devices can be certified for use in functional safety applications according to IEC 61508, providing application developers with the evidence required to demonstrate that the application including the device is also compliant.

IEC 61511 was written and targeted at end-users and provides "best practices" recommendations for users to follow when implementing a Safety Instrumented System (SIS). Let's backup for a few minutes.

Hazards Analysis

The IEC standards prescribe layers of protection to be used. They further state that these layers should (1) be independent of each other, (2) meet a certain reliability level, (3) be capable of being checked or audited, and (4) be specific in their design regarding the level of risk they contain. Protection layers are to be designed to either mitigate (reduce the severity of) a hazardous occurrence or prevent one. Community and plant response plans are mitigation layers, while the basic process design, alarm system, automated SIS and relief devices are prevention type layers of protection. All should be utilized.

The specific layers of protection required will come from the Hazards Analysis. Such an analysis will include an engineering study that reviews the process, electrical, mechanical, safety, instrumentation, and management aspects of the process for those processes with severe risks. The Hazards Analysis will determine if an SIS is required. The purpose of the SIS, a.k.a. safety shutdown or safety interlock, is always to take the process to what is considered a minimal "safe state."

A SIS consists of safety instrumented functions (SIFs) and there can be multiple SIFs within an SIS. Each SIF can contain sensors, signal processing equipment, logic devices, and actuators. A SIF might be an over-pressure control on a vessel. A pressure sensor detects pressure at a level above its normal set-point. Logic then determines that a vent or relief valve should be open to return the vessel to a safe state and the control valve responds to the output from the logic controller and opens, and thereby relieves the pressure buildup until the pressure in the vessel is again detected to be within safe set-point levels.

A SIS can have a single or multiple SIFs. An SIL is assigned to indicate the extent to which a process is expected to perform safely. Process measurement and control equipment and a SIS are not assigned a SIL. The equipment are indicated to be "suitable for use" within a certain SIL environment.

Safety Integrity Level

The Safety Integrity Level (SIL) is established to indicate the acceptable probability of a failure on demand for the safety system to work. The questions the SIL answers are the extent to which a process can be expected to perform safely and to fail safely. They are a measure of safety risk for a given process. The SIL indicates what the tolerable failure rate is for the process. The SIL encompasses the entire process rather than just one specific component. This is determined upon completion of a Hazards Analysis. If the analysis determines that safety protection is sufficient then a SIS is NOT required, otherwise it will be needed.

There are four levels of integrity defined in the standards. Each level represents an order of magnitude of greater risk reduction. The higher the impact a failure can have, the lower the tolerable failure rate will be, and the higher the SIL number will be. For example, SIL 4 has greater risk reduction than SIL 3, which is greater than SIL 2, which is greater than SIL 1.

The effectiveness of an SIS can be described in terms of the probability of it failing to perform its function upon demand to perform. This is called the Probability of Failure on Demand or PFD. The average PFD is used to determine the SIL required. Refer to Table A-2 for the relationship between the PFD_{avg}, SIS availability, risk reduction, the SIL, and potential consequences.

Standards

Refer to the following standards for more information:

- IEC 61508 (issued in 2000)
- IEC 61511 (issued in 2003)
- ANSI/ISA 84 (issued in 2004)

Table A-2   SIL and related parameters

SIL	SIS Availability	PFD$_{avg}$	Reduction of Risk	What Might Happen
4	>99.99%	10^{-5} to <10^{-4}	100,000–10,000	Fatalities in community
3	99.9%	10^{-4} to <10^{-3}	10,000–1,000	Multiple fatalities
2	99–99.9%	10^{-3} to <10^{-2}	1,000–100	Major injuries or a single fatality
1	90–99%	10^{-2} to <10^{-1}	100–10	Minor injuries

Conclusion Regarding Safety Instrumented Systems and Safety Integrity Levels

The effective operation of a group of processes will impact the quality of the product produced and its availability and profitability. Proper safety hazard analysis and implementation of safety systems using SIS and SIL will help ensure that your plant, processes, your neighbors, and perhaps even yourself will be around to benefit from it all.

For critical safe process control of inventory and process vessels, consider instrumentation rated for use in the applicable SIL environment when needed.

3   An Introduction to Hazardous Electrical Locations

Introduction

In North America the target audience for this white paper and the electrical code classification ratings for various types of areas have been based around the Class/Division system. In Europe, this has been based around what is called the multi-tier Zone system. While harmonizing of these two systems is in process in North America, there may be differences between the Canadian and US national electrical codes when it comes to the Zone system. In addition, the Class/Division and Zone systems are similar, but distinct differences do exist.

The fundamentals of hazardous locations are consistent and apply to both the Class/Division and Zone systems. However, the definition of a hazardous environment and the construction of the instrumentation can differ. We illustrate the differences and make a comparison later in this appendix. Refer to Table A-3 and then we will explain some basics from the perspective of the Class/Division system.

Table A-3   Class/Division Versus Zone—Frequency of Occurrence

Condition Frequency/ Material	Class/Division System	Zone System
Continuous/Vapor	Class I, Division 1	Zone 0
Intermittent/Periodic/ Vapor	Class I, Division 1	Zone 1
Abnormal/Vapor	Class I, Division 2	Zone 2
Continuous/Dust	Class II, Division 1	Zone 20
Abnormal/Dust	Class II, Division 2	Zone 21
Continuous/Fibers	Class III, Division 1	Zone 22
Abnormal/Fibers	Class III, Division 2	Zone 22

You can see from Figure A-1 that Class I, Division 1 is broken into Zone 0 and Zone 1 within the European zone system. This is one primary difference between the two systems. The reason is said by Zone proponents to be that many systems in Class I, Division 1 areas are "over-classified." The Zone system allows a vast majority of Class I, Division 1 areas to be classified as Zone 1 or 2.

Electrical instrumentation is applied in a wide variety of industrial environments. Users and the manufacturer of the instrumentation must be aware of the "type" of area the instruments are being installed within. We mean the "type" of area from an electrical code standpoint. This appendix provides an introductory amount of information and guidance regarding area electrical classifications and the approvals that may be required for electrical instrumentation to be installed in the area. However, this is not an exhaustive authority on the subject of electrical locations and the electrical code and other governmental regulations within the country of installation should be consulted. They are the final authority in this matter.

Explosive Environments Defined

The National Electrical Code (NEC) in the United States classifies flammable and combustible liquids on the basis of their likelihood to produce explosive vapors and whether these vapors are present and how often they are present. A fluid considered a *flammable liquid* is one having a flash point below 100°F (38°C) and a vapor pressure not exceeding 40 psia (2.75 bar). A *combustible liquid* is one having a flash point at or above 100°F (38°C). The "flash point"

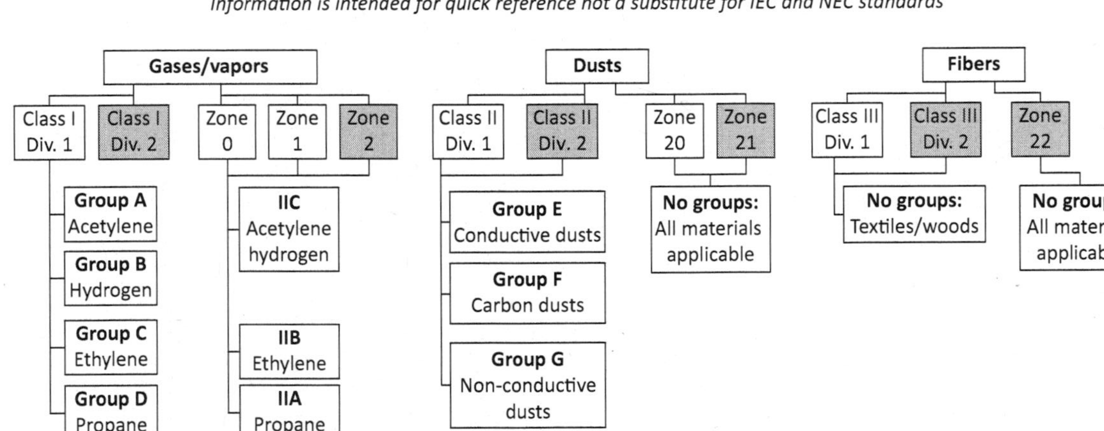

Zone 0	Explosive gas-air mixture is continuously present; comparable to Class I, Div 1.
Zone 1	Explosive gas-air mixture is likely to occur in normal operations; comparable to Class I, Div 1.
Zone 2	Explosive gas-air mixture is not likely to occur; comparable to Class I, Div 2.
Zone 20	Explosive atmosphere resulting from dust continuously present; comparable to Class II, Div 1 applications.
Zone 21	Explosive dust atmosphere is likely to be present occasionally during normal operation; comparable to Class II, Div 2 applications.
Zone 22	Explosive dust atmosphere is unlikely to occur during normal operation, but if it does it is only present for a short period of time; comparable to Class III applications.

Figure A-1   Hazardous location information diagram.

is the minimum temperature at which the fluid will emit vapor in sufficient concentration to become an ignitable mixture with air at the air/liquid interface. Remember, one of the fundamental laws of chemistry is that a fire can exist if the right combination of fuel, heat, and oxygen are present. This is sometimes known as the "fire triangle."

Author note: Consider the Buncefield oil depot accident we discussed earlier in this Appendix. The overfilling of the fuel resulted in several tons of liquid that eventually vaporized into a cloud of explosive vapor and ignited. An excellent analysis of the explosion and vapor cloud progressing is documented in a white paper that can be reviewed for further information on this subject (Davis, 2010).

Examples:

1. *Aviation grade gasoline*: This fluid has a flash point of −50°F (−46°C) and emits explosive vapors at most ambient air conditions. Fluid at Buncefield depot that overflowed its tank was "petrol" or gasoline (automotive), with a flash point of −45°F (−43°C).

2. *Number 1-D diesel fuel*: This fluid has a flash point of 100°F (38°C) and does not emit vapor unless heated to higher temperatures. Facilities handling Number 1-D diesel fuel may not necessarily be classified as hazardous except in hot locations or locations where the fluid could come into contact with hot surfaces.

All flammable gases and liquids heated above their flash points can be ignited. However, the "concentration" also needs to be considered. There is a minimum and maximum concentration level below and above which the propagation of flame will not occur (essentially, mixtures of fuel and oxygen are too lean or too rich). These concentrations are known as the "lower and upper limits of flammability or explosiveness" of the fluid. An increase in temperature or pressure of the fluid will lower the lower limit and raise the upper limit of flammability. A decrease in temperature or pressure has the opposite effect. Some materials have a very wide range of flammability and some materials have a narrow range. Acetylene has a low concentration limit of 2.5% by volume in air and an upper limit of 100%. Gasoline has a narrow range of 1.4–7.6%. Combustible dusts also present serious hazards and have flammable limits, usually called explosion concentrations.

Regarding dusts, the US Chemical Safety Board (CSB) speculates that combustible dusts present a more serious hazard than ignitable vapors because of the lack of awareness as of this writing. The CSB has produced an excellent video on the subject of dust flammability and explosion hazards. Viewing this approximately 30-minute video is highly recommended as it explains the process of dust hazard explosion and illustrates several serious dust explosion and fire accidents. It is available on YouTube or from the media section of the BlueLevel Technologies website at www.blueleveltechnologies.com. The direct YouTube URL is as follows: http://www.youtube.com/watch?v=3d37Ca3E4fA.

With regard to explosive dusts, dust clouds can be so thick that it is impossible to discern objects more than three to five feet away, even when present in minimum explosion concentrations. Usually the concentration is probably below the lower explosion limit if you can see your hand in front of your face. However, danger of fire propagation may still exist even if the concentration is low. This can also produce secondary explosions and can be even more hazardous as the CSB video explains. Please view the CSB video for further information.

Defining Hazardous Areas

The terms hazardous area, hazardous location, and classified location can be used interchangeably. The most common industry term is "hazardous location" and that is what we use here. These terms are also interchangeably used in National Fire Protection Association (NFPA) codes and standards and in the US NEC.

Hazardous locations deal with physical areas of facilities where fire or explosion hazards can or do exist due to the presence of flammable gases/vapors, flammable liquids, combustible dust, or ignitable fibers. The specific type of hazard is defined as an "explosion hazard" in Class I and II locations, and a "fire hazard" in Class III locations.

Class I Locations

Hazardous locations known as or classified as Class I are areas where flammable gases or vapors may be present in the air in quantities sufficient to produce explosive or ignitable mixtures. Class I hazardous locations are broken down into Divisions, and Divisions categorized into Groups. The Divisions indicate how likely or frequently the hazardous gas or vapor exists in the area. This is defined in further detail later in this appendix.

Groups: Class I Groups exist to permit classification of locations based on the explosion characteristics of a reference fluid, regardless of the Division classification as defined later. The NEC definition of Groups within Class I for hazardous gases and vapors is as follows:

- Group A: acetylene
- Group B: hydrogen, or gases/vapors of equivalent hazard
- Group C: ethyl ether vapors, ethylene, or cyclopropane
- Group D: gasoline, hexane, naphtha, benzene, butane, propane, alcohol, acetone, benzyl, lacquer solvent vapors, natural gas

Class II Locations

Whereas Class I deals with explosive gases and vapors, Class II deals with combustible dusts. The dust may be suspended in the air, such as in a cloud, or in a layer resting on electrical equipment or even on the floor, railings, and other structures as a result of a lack of proper dust collection equipment. In 1979 the National Materials Advisory Board reported to OSHA that for most dusts the ignition temperature of a dust layer is usually lower than for a dust cloud. In those cases where the dust cloud ignition temperature was lower than the layered temperature it was apparent that either (1) there was a change in state from a dust to a gas, (2) the material in the cloud was somewhat different than in the layered form, or (3) the difference in temperatures between layered and cloud were not really significant.

However, as the CSB video referenced before clearly illustrates, dust layers can become airborne as a result of cleaning and even initial explosions, and then provide fuel for subsequent explosions and fire propagation. Class II locations are broken down into Divisions, which represent the likelihood of a dust explosion if ignition sources are not carefully controlled.

Groups: As with Class I, Class II Groups exist to permit classification of locations based on the explosion characteristics of the material. For Class II locations there are three Groups:

- Group E: metal dusts (aluminum, magnesium, and their alloys) and other combustible dusts whose particle size, abrasiveness, and conductivity present similar hazards
- Group F: carbonaceous dusts (carbon black, charcoal, coal, or coke) that have more than 8% total entrapped volatiles, or dusts sensitized by other materials so they present an explosion hazard
- Group G: combustible dusts not included in groups E and F

Instrumentation/Equipment Design

Class I locations are a more serious hazard to protect against, from the standpoint of instrument design. Therefore, we will only look at the impact this Class has upon instrument design. In the design and manufacture of instrumentation suitable for use in a Class I environment there are two approaches; (1) the most common is to build the instrument to be "explosion proof" or capable of withstanding an explosion, (2) the second approach is to build an instrument that is designed with increased safety in that the device is "intrinsically safe." An intrinsically safe device is one that will never generate the minimum energy necessary to ignite an explosive atmosphere. With an explosion proof instrument it is assumed the flammable mixture will enter the enclosure and that the enclosure will contain any source of ignition from escaping the instrument housing even if an explosion occurs within the instrument enclosure. The latter is the most common approach.

To provide explosion proof protection the objective of the instrumentation manufacturer in designing the instrument is to build the enclosure strong enough to withstand the anticipated explosion, and to prevent the resulting flames and hot gases from igniting the surrounding atmosphere outside the instrumentation enclosure.

To assist the instrumentation manufacturer during design, a "maximum experimental safety gap" (MESG) is defined for each Group of materials. This gap is a maximum dimension through which an ignition inside the enclosure will not propagate flame to a mixture of flammable material outside the enclosure. The joints in an enclosure are considered to be the "flamepath" and these are evaluated and properly designed in addition to the strength of the enclosure itself. Each material or fluid is classified into a Group based on its value for the MESG.

The maximum allowable joint gap depends on the type of joint (plain, stepped, or threaded), the Group classification, and the volume of the instrument enclosure. Threaded joints are also concerned with the width of the sealing joint area, the thread itself (≤ 20 threads per inch), and having a minimum thread engagement (5 minimum). The detail for the design of instrumentation regarding the MESG is laid out within the applicable country standards such as CSA (Canadian Standards Association) and UL (Underwriters Laboratory). Refer to standards documents for further information.

Instrument Surface Temperature

The ignition temperature of a flammable material is critical in determining the acceptability of instrumentation operating in hazardous locations. The reason for this is that the external surface of the enclosure itself can act as an ignition source. This assumes the surface temperature of the external area of the enclosure can rise to a point that exceeds the ignition temperature of the flammable material. This is why the ignition temperature of the material is important. High surface temperatures can exist not just because of a fault (overheating) within the instrument but also as a result of normal operation.

To assist in understanding whether an instrument may produce surface temperatures greater than the ignition temperature of the material, instruments destined for installation in hazardous areas are marked with a code that represents the maximum external surface temperature that can be expected with the instrument.

Equipment cannot be used in hazardous areas when the maximum surface temperature is greater than the ignition temperature of the gases and vapors present. The allowable ambient temperature may be limited to comply with the temperature rating of the device. External surface temperature codes and their related temperatures can be found in Table A-4.

Table A-4   Maximum allowable surface temperature on external surface of instrument enclosure

Maximum Temperature	Identification Number
450°C (842°F)	T1
300°C (572°F)	T2
280°C (536°F)	T2A
260°C (500°F)	T2B
230°C (446°F)	T2C
215°C (419°F)	T2D
200°C (392°F)	T3
180°C (356°F)	T3A
165°C (329°F)	T3B
160°C (320°F)	T3C
135°C (275°F)	T4
120°C (248°F)	T4A
100°C (212°F)	T5
85°C (185°F)	T6

Classification "Divisions"

In the Class system within the US NEC, Divisions are used to identify the likelihood or frequency of a flammable or ignitable mixture being present. Divisions within Class I are as follows.

Class I, Division 1: Ignitable concentrations of flammable gases or vapors can exist under normal operating conditions. These ignitable concentrations may exist frequently because of maintenance or leakage. Ignitable concentrations can exist because of faulty operation of equipment or processes resulting in failure of electronic equipment.

Instrument enclosures approved for Class I, Division 1 locations include those that provide protection by being explosion proof (sometimes referred to as flameproof) or being purged/pressurized. Explosion proof enclosures are the most common method of protection for instrumentation in these hazardous locations. The instrument enclosure is capable of withstanding an internal explosion of a specified gas or vapor and preventing the ignition of the same gas or vapor in the surrounding ambient environment outside the instrument enclosure as a result of any sparks, flashes, etc. from the explosion inside the enclosure. The external surface temperature of the instrument enclosure is such that the surrounding gases or vapors will not be ignited.

An alternate method to designing an explosion proof enclosure meeting the Class I, Division 1 requirements for a specific Group is to purge and pressurize the instrument enclosure. This approach eliminates the presence of flammable vapors from within the enclosure. This allows enclosures that are not otherwise acceptable for use in hazardous locations to be useable when this technique is implemented.

How does purging and pressurizing an enclosure work? Purging supplies the instrument enclosure with a protective gas (such as nitrogen) at enough flow rate and pressure to lower the concentration of any flammable gas that might exist within the enclosure to an acceptable level according to the standards. There are three types of purging. Type "X" purging lowers the enclosure classification from Division 1 to Ordinary (general purpose). Type "Y" purging will reduce the enclosure requirement from Division 1 to Division 2. Type "Z" purging reduces the enclosure requirement from Division 2 to Ordinary (general purpose).

Enclosures for Ordinary locations (general purpose suitable for non-hazardous locations) are also acceptable for Class I, Division 1 locations if the electronics within them are intrinsically safe. In this situation the proper electrical barriers must be installed on any instrument I/O circuits. Intrinsically safe electronics have low energy circuitry that keeps energy levels below those needed to ignite any flammable gas or vapor that might be present. This intrinsically safe approach takes into account the energy available during normal and abnormal operating conditions.

Class I, Division 2: Instruments used in these Division 2 areas are in locations where flammable fluids or gases exist, but in which it is normally confined and can only escape due to accidental bursting of containers or during abnormal operation. Class I, Division 2 locations can also be where the accumulation of ignitable gases or vapors is prevented by mechanical ventilation and can only occur through the failure or abnormal operation of equipment. These locations can be adjacent to Class I, Division 1.

Instrument enclosures approved for use in Class I, Division 1 areas can also be used in Class I, Division 2 locations. Therefore, the same approach to protection and enclosure design can be used. Ordinary location (general purpose) enclosures may be acceptable only if any current interrupting contacts are hermetically sealed against the entrance of any gas or vapor, or if the electronics are non-incendive and intrinsically safe.

The term "non-incendive" only applies to Division 2 locations. It applies to the circuits used in the instrument. A non-incendive circuit is one in which any arc or heat produced during normal operating conditions is not able to ignite a flammable gas/vapor or dust mixture. The use of non-incendive circuits is not allowed in Division 1 areas because they only consider normal operating conditions and not abnormal conditions.

Class II, Division 1: In these areas combustible dust is in the air under normal conditions in sufficient concentration to be able to produce explosive or ignitable mixtures. In these locations the failure or abnormal operation of machinery can cause such mixtures and is a source of ignition. Combustible dusts of an electrically conductive nature are also present in these hazardous areas.

Class II, Division 2: These locations have combustible dust; however, the dust is not present normally in quantities that are sufficient to produce a hazard. Dust accumulation is not sufficient to interfere with electrical equipment. Dust can be in suspension within the air as a result of infrequent equipment breakdown.

Class III: Class III areas are hazardous because easily ignitable fibers are present. However, these fibers are not likely to be in the air in such quantities as would be sufficient to produce any ignitable mixture. There are no Group subdivisions for Class III locations but there are Divisions based on the likelihood of the fiber material being present.

Class III, Division 1: These areas are where the ignitable fibers are handled, manufactured, or used.

Class III, Division 2: These areas are where the ignitable fibers are stored or handled not in the process of manufacture.

The IEC (International Electrotechnical Commission)

The IEC classification system differs from the NEC system though both exist for the same reason. The IEC system differentiates between mine and surface areas by classification I being the former and II being the latter. The IEC system is based on the IEC standard number 60079-2 and uses a Zone system rather than Class and Divisions. There are three zones (0, 1, and 2). Zone 0 corresponds to the most hazardous part of Division 1. Zone 1 is the remainder of Division 1, and Zone 2 is the same as Division 2 (see Figure A-1).

The lettering under the IEC system for dusts (NEC Class II) is Zones 20, 21, and 22. Zone 20 is when a hazardous dust cloud is "likely to be present on a continuous basis or for long periods of time," Zone 21 for when the hazardous dust is "likely to be present occasionally in normal operation," and Zone 22 is when hazardous dust is "unlikely to occur during normal operation, but if it does it is only present for a short period."

Testing Agencies

Instrumentation and other electrical equipment being used within any of the above hazardous areas should have an approval or certification for the intended area from a third-party testing agency. The primary testing laboratories in North America that can test to standards in the United States and Canada and render an approved or certified status for the equipment include:

- Canadian Standards Association
- LabTest Certification, Inc.
- Factory Mutual
- Underwriters Laboratories

The above agencies are approved to perform testing for equipment to determine compliance to the North American standards. Most have bilateral arrangements allowing for testing to both UL (United States) and CSA (Canada) standards and approvals or certifications, meaning that each is accepted almost everywhere in North America no matter where the testing lab is located.

4 Conclusion

Level monitoring is a critical application that ensures safety by detecting high and low material levels, and ensures profitability by monitoring and managing the inventory of materials in tanks, bins, and silos. A wide variety of technologies and brands exist. It can sometimes be a complex task to determine which product to use. With regard to safety of your required electrical classifications, if you have not spent decades in the industry you may need an introduction to hazardous areas. We hope we have been of service to you here. Please do not hesitate to give us a call with any questions..

5   Appendix for Hazardous Locations

Enclosure NEMA Ratings

Enclosure Type	Protection Against the Following Environmental Conditions
1	Indoor use.
2	Indoor use only. Limited amount of falling water.
3R	Drip-tight. Outdoor use, undamaged by a formation of ice.
3	Rainproof. Same as 3R plus windblown dust.
3S	Rain-tight, Dust-tight. Same as 3 with external mechanisms remaining operable when ice laden.
4	Outdoor use, splashing water, windblown dust, hose-directed water, undamaged by ice formation.
4X	Rain-tight, watertight. Same as 4 plus corrosion resistant.
5	Indoor use to provide a degree of protection against settling airborne dust, falling dirt, and dripping non-corrosive liquids.
6	Drip-tight, dust-tight. Same as 3R plus entry of water during temporary submersion at limited depth.
6P	Rain-tight, watertight. Same as 3R plus entry of water during prolonged submersion at limited depth.
7	Indoor use, in hazardous locations classified as Class I, Division 1, Groups A, B, C, or D.
8	Indoor or outdoor use, in hazardous locations classified as Class I, Division 1, Groups A, B, C, or D.
9	Indoor use, in hazardous locations classified as Class II, Division 1, Groups E, F, or G.
12, 12K	Indoor use, dust, dripping non-corrosive liquids, drip-tight, dust-tight.

Enclosure Ingress Protection

IP X Y

Protection Against Solids (X)	Protection Against Liquids (Y)
0: No protection	**0**: No protection
1: Objects > 50 mm	**1**: Vertically dripping water
2: Objects > 12 mm	**2**: Angled dripping water -75 to 90°C
3: Objects > 2.5 mm	**3**: Sprayed water
4: Objects > 1.0 mm	**4**: Splashed water
5: Dust protected	**5**: Water jets
6: Dust-tight	**6**: Heavy seas
	7: Effects of immersion
	8: Indefinite immersion

A device listed as IP 66 indicates that the device is dust-tight and will be protected against heavy seas during operation. The IP designation sets the degree of environmental protection that a given device has. The IP designation describes how well the device resists both solid and liquid ingress. The applicable IEC standard is IEC 60529.

Reference

Davis, Scott G., Peter Hinze, Olav R. Hansen, and Kees van Wingerden. 2010. 2005 Buncefield vapor cloud explosion: Unraveling the mystery of the blast. GexCon US, Bethesda, Maryland.

Appendix B

Bulk Solids Flow

This appendix provides information regarding a topic of great importance in many level measurement and detection applications. The ability of the bulk solid material to flow when discharging from a bin, silo, or other vessel can impact the overall success of either a level control or inventory monitoring application. Bridging, ratholes, and other difficult flow problems where material will not properly discharge occur for a variety of reasons and should be resolved to ensure proper level control and inventory measurement of a powder and granular material. Prevention seems the best approach and planning for this prior to the initial design, purchase, and installation of the vessels may avoid these problems. However, the reality is that dealing with the aftermath is most common. This appendix provides information regarding bulk solids flow as an educational tool so the reader will understand how flow problems are created and may be resolved.

There are essentially three basic methods to resolve flow problems associated with powders and bulk solids within bins. This includes vessel design, vibration, and aeration. Most approaches will fall into one of these three methods.

Vessel Design: The first method, vessel design, purports to prevent the flow problem by converting the material flow from funnel flow to mass flow. One such example is illustrated in an October 2003 article appearing in *POWER* magazine that describes a particularly difficult and dangerous flow problem that resulted in an explosion at the facility in June 1991 (Dantoin et al., 2003). Subsequently the company made several changes including changing its coal hoppers so that the material flowed as mass flow. To achieve this, the hopper walls of the cone section were changed to a much steeper angle, the discharge size was increased, and a new lining was installed in place of the existing Gunite lining. These changes were made on two coalbunkers at a very high price, but the results have been good: no fires or explosions since, as reported by the article.

Vibration: The second method, vibration, can also be an effective means of dealing with flow problems. However, unlike vessel design the use of industrial vibrators does not treat the problem at its root, it only deals with the symptom and, in some regards defies logic. When you vibrate a powder in a container you actually pack the material, the exact opposite of what you would want to do to resolve the flow problem, right? That's what I would think too, but the use of high levels of vibration energy transmitted through the vessel wall into the material serves to break loose the material particles from the walls of the vessel and each other. Many flow problems are a result of adhesion (material adhering to vessel walls) and cohesion (particles packing together, clinging to one another and forming lumps, clumps, and a solid mass that is difficult to flow). Vibration energy, enough of it, can dislodge many clogs of this type and thereby promote material flow. Sizing and selecting the correct industrial vibrator is important (Ruggio, 2003).

Aeration: The third method, aeration, like vibration, also does not deliver a true solution of the cause of the flow problem. Aeration uses air pressure to separate bulk solid particles from each other and from vessel walls. As we said before, adhesion and cohesion are culprits in flow problems. The air injected into the material at the location of the flow problems serves to lift and separate material particles from each other and the vessel walls thereby promoting flow. Where vibration may typically use a single industrial vibrator to resolve the flow problem, but not always, the use of bin aerators almost always requires a number of aerators throughout the vessel section where the problem is. An example of the common use of aerators to resolve material flow problems is in the cone section of silos that contain cement powder or fly ash within a concrete batching plant.

Technical Paper: The information that follows is a technical paper titled "Flow Properties of Powder and Bulk Solids" authored by Dr. Dietmar Schulze. It is reprinted with permission in its entirety. It provides more in-depth discussion of how flow problems can be determined, how they create real-world problems, and how they may be resolved.

References

Dantoin, Bruce, Rod Hossfeld, and Kerry McAtee. 2003. Converting from funnel flow to mass flow. *POWER* (October).

Ruggio, David. 2003. Demystifying bin vibrator selection. *Powder Bulk Solids* (September).

Flow Properties of Powders and Bulk Solids

Dr. Dietmar Schulze,
Schüttgutmesstechnik,
Am Forst 20,
D-38302 Wolfenbüttel,
Germany
Tel. ++49 (0) 5331 935 490;
Fax ++49 (0) 5331 978 001;
mail@dietmar-schulze.de;
www.dietmar-schulze.de

In order to compare and optimize powders regarding flowability and to design powder handling equipment like silos, feeders, and flow promoting devices, it is necessary to know the mechanical properties—the so-called flow properties. In the present paper it is outlined which physical parameters describe the flow properties of a powder or a bulk solid, and how these parameters are determined experimentally.

1   Introduction

Knowledge of the flow properties of a powder or a bulk solid is necessary to design silos and other bulk solid handling equipment so that no flow problems (flow obstructions, segregation, irregular flow, flooding, etc.) occur. Furthermore, quantitative information regarding flowability of bulk products is required, e.g., as part of comparative tests (e.g., effect of flow agents or other additions on flow behavior) and quality control. The flow properties depend on several parameters, e.g.,

- particle size distribution,
- particle shape,

- chemical composition of the particles,
- moisture,
- temperature.

It is not possible to determine theoretically the flow behaviour of bulk solids in dependence of all of these parameters. Even if this were possible, the expense for the determination of all parameters of influence would be very high. Thus it is necessary, and also simpler, to determine the flow properties in appropriate testing devices.

The present paper deals with all kinds of particulate solids, which are also called bulk solids, powders, or granulates. In the following the general expression "bulk solid" is used for all these products.

2   Stresses in bulk solids

Figure 1 shows a bulk solid element in a container (assumptions: infinite filling height, frictionless internal walls). In the vertical direction, positive normal stress ($\sigma_v > 0$, compressive stress) is exerted on the bulk solid.

If the bulk solid were to behave like a Newtonian fluid, the stresses in the horizontal and vertical direction (and in all other directions) would be of equal magnitude. In reality the behaviour of a bulk solid is quite different from that of a fluid, so that the assumption of analogies is often misleading.

Within the bulk solid (Figure 1) the horizontal stress, σ_h, is a result of the vertical stress, σ_v, where the resulting horizontal stress is less than the vertical stress exerted on the bulk solid from the top. The ratio of horizontal stress to vertical stress is the stress ratio, K (also known as λ).

$$K = \sigma_h / \sigma_v \tag{1}$$

Typical values of K are between 0.3 and 0.6 [15].

It follows that—in analogy to solids—in a bulk solid different stresses can be found in different cutting planes. Stresses in cutting planes other than the vertical and the horizontal can be analyzed using a simple equilibrium of forces:

No shear stresses τ are exerted on the top or bottom surface of the bulk solid element in Figure 1; i.e., the shear stresses in these planes are equal to

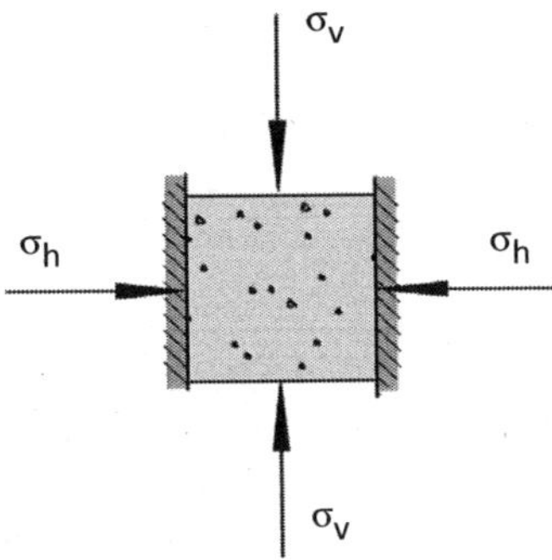

Figure 1   Element of bulk solid.

zero. No shear stresses are acting at the lateral walls, since the lateral walls were assumed as frictionless. Thus only the normal stresses shown are acting on the bulk solid from outside. Using a simple equilibrium of forces at a volume element with triangular cross-section cut from the bulk solid element shown in Figure 1 (Figure 2, on the left), the normal stress, σ_a, and the shear stress, τ_a, acting on a plane inclined by an arbitrary angle a, can be calculated. After some mathematical transformations, which need not be considered here, it follows that:

$$\sigma_a = \frac{\sigma_v + \sigma_h}{2} + \frac{\sigma_v - \sigma_h}{2}\cos(2a) \tag{2}$$

$$\tau_a = \frac{\sigma_v - \sigma_h}{2}\sin(2a) \tag{3}$$

The pair of values (σ_a, τ_a), which are to be calculated according to equations (2) and (3) for all possible angles a, can be plotted in a σ,τ-diagram (normal stress, shear stress diagram); see Figure 2 on the right. If one joins all plotted pairs of values, a circle emerges; i.e., all calculated pairs of values form a circle in the σ,τ-diagram.

This circle is called "the Mohr stress circle." Its centre is located at $\sigma_m = (\sigma_v + \sigma_h)/2$ and $\tau_m = 0$. The radius of the circle is $\sigma_m = (\sigma_v - \sigma_h)/2$. The Mohr stress circle represents the stresses on all cutting planes at arbitrary inclination angles a, i.e., in all possible cutting planes within a bulk solid element.

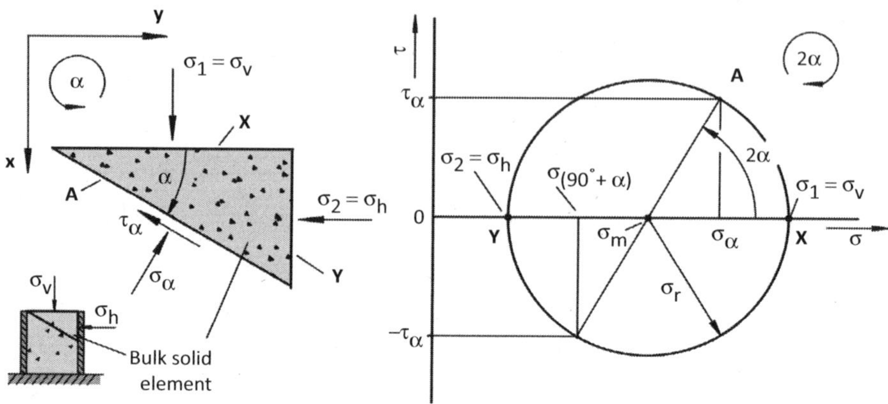

Figure 2   Force equilibrium on an element of bulk solid, the Mohr stress circle.

Since the centre of the Mohr stress circle is always located on the σ-axis, each Mohr stress circle has two points of intersection with the σ-axis. The normal stresses defined through these points of intersection are called the principal stresses, whereby the larger principal stress—the major principal stress—is designated as σ_1 and the smaller principal stress—the minor principal stress—is designated as σ_2. If both principal stresses are given, the Mohr stress circle is well defined.

In the example of Figure 1 both the horizontal and the vertical plane are free from shear stresses ($\tau = 0$) and are thus principal stress planes. In this case the vertical stress, σ_v, which is greater than the horizontal stress, σ_h, is the major principal stress, σ_1, and the horizontal stress, σ_h, is the minor principal stress, σ_2. An important qualitative result of the above analysis is that shear stresses can occur in bulk solids at rest. This is impossible for a Newtonian fluid at rest (in contrast to Newtonian fluids, bulk solids can have sloped surfaces even at rest). Therefore, a representation of the stresses (fluids: pressures) in different cutting planes of a Newtonian fluid at rest in a σ,τ-diagram would yield a stress circle with the radius zero (equation (3) with $\sigma_h = \sigma_v$ yields $\tau_a = 0$).

From the explanation above it follows that the state of stress in a bulk solid cannot be completely described by only a single numerical value. Depending on the actual load acting on a bulk solid element, the corresponding Mohr stress circle can have a smaller or a larger radius, a centre at a lesser or greater normal stress, and hence also different principal stresses, σ_1 and σ_2.

In principle, at a given major principal stress, σ_1, stress circles with different values for the lowest principal stress, σ_2, are imaginable. Therefore, a stress circle is defined clearly only if at least two numerical values are given, i.e., σ_1 and σ_2.

In summary, the following can be stated with regard to the stresses acting in bulk solids:

- A bulk solid can transmit shear stresses even if it is at rest.
- In different cutting planes different stresses are acting.
- Stress conditions can be represented with Mohr stress circles.

3   Adhesive forces

The flowability of a bulk solid depends on the adhesive forces between individual particles. Different mechanisms create adhesive forces [5]. With fine-grained, dry bulk solids, adhesive forces due to van der Waals interactions play the essential role. With moist bulk solids, liquid bridges between the particles usually are most important. Liquid bridges are formed by small regions of liquid in the contact area of particles, in which due to surface tension effects a low capillary pressure prevails.

Both types of adhesive forces described above are dependent on the distance between particles and on particle size.

Some bulk solids continue to gain strength if stored at rest under compressive stress for a longer time interval. This effect is called time consolidation. The reasons for time consolidation are also to be found in the effects of adhesive forces. Possible mechanisms are:

- Solid bridges due to solid crystallizing when drying moist bulk solids, where the moisture is a solution of a solid and a solvent [5] (e.g., sand and salt water).
- Solid bridges from the particle material itself, e.g., after some material at the contact points has been dissolved by moisture [5] (e.g., crystal sugars with slight dampness).
- Bridges due to sintering during storage of the bulk solid at temperatures not much lower than the melting temperature [5]. This can appear e.g., at ambient temperature during the storage of plastics with low melting points.

- Plastic deformation at the particle contacts, which leads to an increase in the adhesive forces through approach of the particles and enlargement of the contact areas.
- Chemical processes (chemical reactions at the particle contacts).
- Biological processes (e.g., due to fungal growth on biologically active ingredients).

Whether a bulk solid flows well or poorly depends on the relationship of the adhesive forces to the other forces acting on the bulk solid. It can be shown that the influence of adhesive forces on flow behaviour increases with decreasing particle size. Thus, as a rule, a bulk solid flows more poorly with decreasing particle size. Fine-grained bulk solids with moderate or poor flow behaviour due to adhesive forces are called cohesive bulk solids.

If particles are pressed against each other by external forces, the compressive force acting between the particles increases. Thereby large stresses prevail (locally) at the particles' contact points, because the contact points are very small. This leads to plastic deformation of the particles in the contact area, so that the contact areas increase and the particles approach each other. Thereby the adhesive forces increase. Thus a compressive force acting from outside on a bulk solid element can increase the adhesive forces. This mechanism is used, e.g., in the production of tablets or briquettes.

The dependence of the adhesive forces between the particles on external forces exerted on a bulk solid is characteristic of bulk solids, especially for cohesive bulk solids. Therefore an evaluation of bulk solids behaviour must always take into consideration the forces or stresses previously acting on the bulk solid, the stress history. The stress history includes, for example, the consolidation stress exerted on a bulk solid, leading to certain adhesive forces and hence to a certain strength of the bulk solid (e.g., the strength of a tablet is dependent on the maximum consolidation stress at tabletting).

4   Flowability

4.1   Uniaxial compression test

The phrase "good flow behaviour" usually means that a bulk solid flows easily, i.e., it does not consolidate much and flows out of a silo or a hopper due to the force of gravity alone and no flow promoting devices are required. Products are "poorly flowing" if they experience flow obstructions or consolidate during storage or transport. In contrast to these qualitative statements, a quantitative statement on flowability is possible only if one uses an objective characteristic value that takes into account those physical characteristics of the bulk solid that are responsible for its flow behaviour.

"Flowing" means that a bulk solid is deformed plastically due to the loads acting on it (e.g., failure of a previously consolidated bulk solid sample). The magnitude of the load necessary for flow is a measure of flowability. This will be demonstrated first with the uniaxial compression test. Figure 3 shows a hollow cylinder filled with a fine-grained bulk solid (cross-sectional area A; internal wall of the hollow cylinder assumed as frictionless). The bulk solid is loaded by the stress σ_1—the consolidation stress—in the vertical direction. The more the volume of the bulk solid specimen is reduced, the more compressible the bulk solid is.

In addition to the increase in bulk density from consolidation stress, one will observe also an increase in strength of the bulk solid specimen. Hence, the bulk solid is both consolidated and compressed through the effect of the consolidation stress.

After consolidation, the bulk solid specimen is relieved of the consolidation stress, σ_1, and the hollow cylinder is removed. If subsequently the consolidated cylindrical bulk solid specimen is loaded with an increasing vertical compressive stress, the specimen will break (fail) at a certain stress. The stress causing failure is called compressive strength or unconfined yield strength, σ_c (another common designation is f_c).

In bulk solids technology one calls the failure "incipient flow," because at failure the consolidated bulk solid specimen starts to flow. Thereby the bulk solid dilates somewhat in the region of the surface of the fracture, since the distances between individual particles increase. Therefore incipient flow is

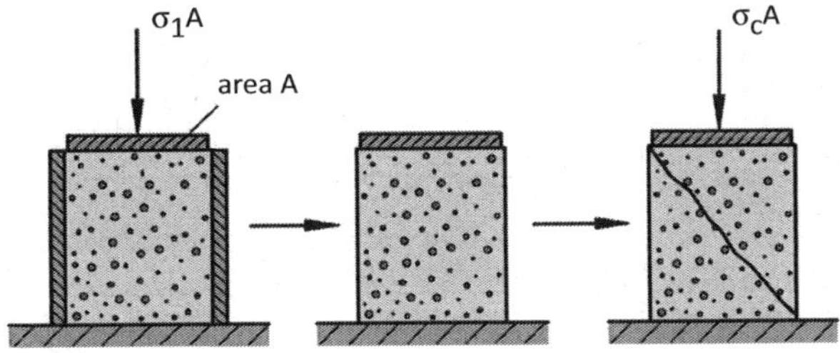

Figure 3   Uniaxial compression test.

plastic deformation with decrease of bulk density. Since the bulk solid fails only at a sufficiently large vertical stress, which is equal to the compressive strength, there must exist a material-specific yield limit for the bulk solid. Only when this yield limit is reached does the bulk solid start to flow.

The yield limits of many materials (e.g., metals) are material-dependent and are listed in tables. However, the yield limit of a bulk solid is dependent also on its stress history, i.e., previous consolidation: The greater the consolidation stress, σ_1, the greater the bulk density, ρ_b, and unconfined yield strength, σ_c.

Uniaxial compression tests (Figure 3) conducted at different consolidation stresses, σ_1, lead to different pairs of values (σ_c, σ_1) and (ρ_b, σ_1). Plotting these pairs of values as points in a σ_c, σ_1-diagram and a ρ_b, σ_1-diagram, respectively, and drawing in each diagram a curve through these points, usually results in curves like those for product A in Figure 4, where bulk density, ρ_b, and unconfined yield strength, σ_c, typically increase with consolidation stress, σ_1. Very rarely a progressive slope like in the left part of curve B is observed. The curve $\sigma_c(\sigma_1)$ is called the flow function.

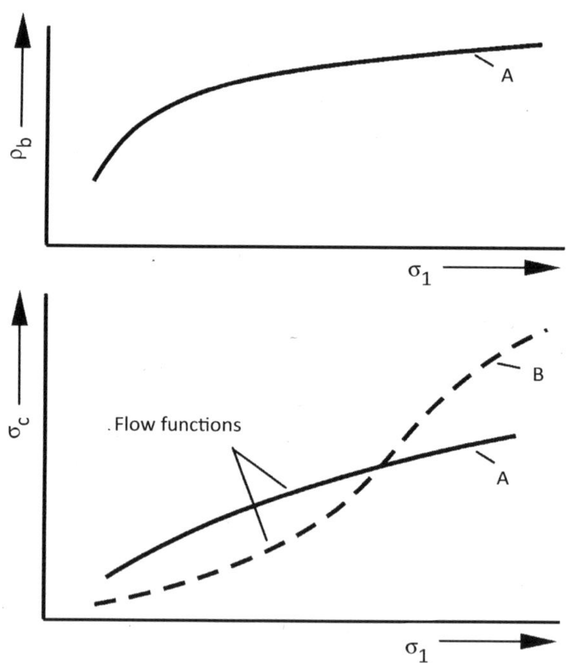

Figure 4 Bulk density, ρ_b, and unconfined yield strength, σ_c, vs. consolidation stress, σ_1.

4.2   Time consolidation (caking)

Some bulk solids increase in strength if they are stored for a longer time at rest under a compressive stress (e.g., in a silo or an intermediate bulk container). This effect is called time consolidation or caking. Time consolidation can be determined with the test shown in Figure 3, in order, e.g., to simulate long-term storage in a silo. For this one loads the specimen with consolidation stress, σ_1, not only for a short moment, but for a defined period of time, t_1. Then the unconfined yield strength is determined following the principle explained above (Figure 3).

Figure 5 shows the flow function $\sigma_c(\sigma_1)$ of product A as previously shown in Figure 4 (unconfined yield strength without influence of time consolidation, i.e., for a storage period $t = 0$). Additionally, examples of curves $\sigma_c(\sigma_1)$ for storage periods $t > 0$ (curves A_1, A_2) are drawn. The curves $\sigma_c(\sigma_1)$ for the storage periods $t > 0$ are called time flow functions. Here each curve emerges from the connection of several pairs of values (σ_c, σ_1), which were measured at identical storage periods, t, but at different consolidation stresses, σ_1.

For the example of bulk solid A, the unconfined yield strength, σ_c, increases with increasing storage time. This result is true for many bulk solids, but not for all. There are bulk solids which undergo no or only very slight consolidation over time; i.e., σ_c does not increase, or increases only very slightly with increasing storage period, t (e.g., dry quartz sand). Other bulk solids undergo a large increase in unconfined yield strength after storage periods of only a few hours, whereas after longer storage periods their unconfined yield strength

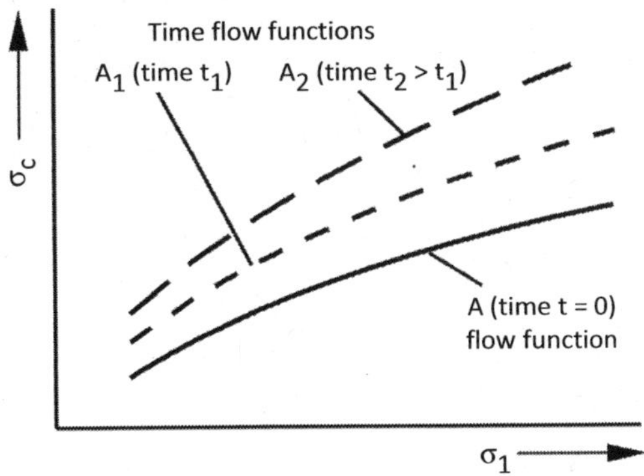

Figure 5   Flow function and time flow functions for two different storage times t_1 and $t_2 > t_1$.

does not increase further. These differences are due to the different physical, chemical, or biological effects that are the causes of consolidation over time, e.g., chemical processes, crystallizations between the particles, enlargement of the contact areas through plastic deformation, capillary condensation, or biological processes such as fungal growth (see Section 3).

With measurement of time consolidation, a "time-lapse effect" is not realizable; i.e., one must store a bulk solid specimen at the consolidation stress, σ_1, for exactly that period of time for which one would like to get data on time consolidation. Without such a test no quantitative statement can be made regarding time consolidation.

4.3   Yield limit and Mohr stress circles

The uniaxial compression test presented in Figure 3 is shown below in a σ,τ-diagram (Figure 6). If one neglects the force of gravity of the bulk solid specimen and assumes that no friction is acting between the wall of the hollow cylinder and the bulk solid, both vertical stress, σ_v, as well as horizontal stress, σ_h, are constant within the entire bulk solid specimen. Therefore at each position in the bulk solid sample the state of stress, which can be represented by a Mohr stress circle, is identical.

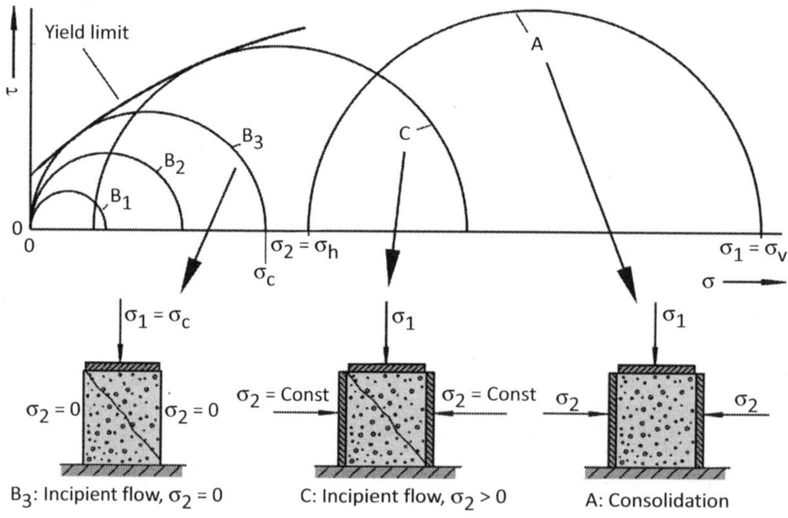

Figure 6   Measurement of unconfined yield strength in a σ,τ-diagram.

During consolidation the vertical normal stress, σ_1, acts on the top of the bulk solid specimen. Perpendicular to the vertical stress the lesser, horizontal stress prevails according to stress ratio K (see Section 2). Neither at top nor at bottom of the specimen, nor at the internal wall of the hollow cylinder, which is assumed as frictionless, will shear stresses be found; i.e., $\tau = 0$. The pairs of values (σ, τ) for vertical and horizontal cutting planes within the bulk solid specimen are plotted in the σ, τ-diagram (Figure 6). Both points are located on the σ-axis because $\tau = 0$. The Mohr stress circle A, which describes the stresses in the bulk solid sample at consolidation, is thus well defined (because each stress circle has exactly two intersections with the σ-axis).

Since in the vertical plane the shear stress is zero, the vertical stress is identical with the major principal stress, σ_1 (see Section 2: Principal stresses are the normal stresses in those planes in which the shear stresses are equal to zero). The major principal stress, σ_1, is equal to the vertical stress, σ_v, and the minor principal stress, σ_2, is equal to the horizontal stress, σ_h.

In the second part of the test shown in Figure 3, the specimen is loaded with increasing vertical stress after it has been relieved of the consolidation stress and the hollow cylinder has been removed. The vertical stress and horizontal stress are principal stresses. The horizontal stress is independent of the vertical load equal to zero, since the lateral surface of the specimen is uncovered and not loaded.

During the increasing vertical load in the second part of the test, the stress states at different load steps are represented by stress circles with increasing diameter (stress circles B_1, B_2, B_3 in Figure 6). The lesser principal stress, which is equal to the horizontal stress, is equal to zero at all stress circles.

At failure of the specimen the Mohr stress circle B_3 represents the stresses in the bulk solid sample. Since the load corresponding to this Mohr stress circle causes incipient flow of the specimen, the yield limit of the bulk solid must have been attained in one cutting plane of the specimen. Thus, Mohr stress circle B_3 must reach the yield limit in the σ, τ-diagram. In Figure 6 a possible yield limit is shown. The real course of the yield limit cannot be determined with only the uniaxial compression test.

The Mohr stress circles B_1 and B_2, which are completely below the yield limit, cause only an elastic deformation of the bulk solid specimen, but no failure and/or flow. Stress circles larger than stress circle B_3, and thus partly above the yield limit, are not possible: The specimen would already be flowing when the Mohr stress circle reaches the yield limit (failure), so that no larger load could be exerted on the specimen.

If, during the second part of the experiment shown in Figure 3 (measurement of compressive strength), one were to apply also a constant horizontal stress $\sigma_h > 0$ on the specimen (in addition to the vertical stress, σ_v), one would likewise find stress circles that indicate failure of the specimen and reach the yield limit (e.g., stress circle C in Figure 6). Thus the yield limit is the envelope of all stress circles that indicate failure of a bulk solid sample.

4.4   Numerical characterization of flowability

Flowability of a bulk solid is characterized mainly by its unconfined yield strength, σ_c, in dependence on consolidation stress, σ_1, and storage period, t. Usually the ratio ff_c of consolidation stress, σ_1, to unconfined yield strength, σ_c, is used to characterize flowability numerically:

$$ff_c = \sigma_1 / \sigma_c \tag{4}$$

The larger ff_c is, i.e., the smaller the ratio of the unconfined yield strength, σ_c, to the consolidation stress, σ_1, the better a bulk solid flows. Similar to the classification used by Jenike [1], one can define flow behaviour as follows:

$$
\begin{array}{ll}
ff_c < 1 & \text{not flowing} \\
1 < ff_c < 2 & \text{very cohesive} \\
2 < ff_c < 4 & \text{cohesive} \\
4 < ff_c < 10 & \text{easy-flowing} \\
10 < ff_c & \text{free-flowing}
\end{array}
$$

In Figure 7 the flow function A taken from the σ_c,σ_1-diagram in Figure 4 is shown. Additionally, the boundaries of the ranges of the classifications listed above are shown as straight lines, each representing a constant value of flowability, ff_c. This diagram clearly shows that the flowability, ff_c, of a specific bulk solid is dependent on the consolidation stress, σ_1 (in most cases ff_c increases with σ_1 as with bulk solid A). Therefore, with each consolidation stress at which σ_c and thus ff_c were determined, one obtains a different value of flowability: The flowability of a bulk solid depends on the stress level (= consolidation stress); thus for most bulk solids one will obtain a larger value of flowability (= better flowability) at a greater consolidation stress. For most bulk solids one will find a (possibly extremely low) consolidation stress at which the bulk solid flows poorly. Because of the dependence

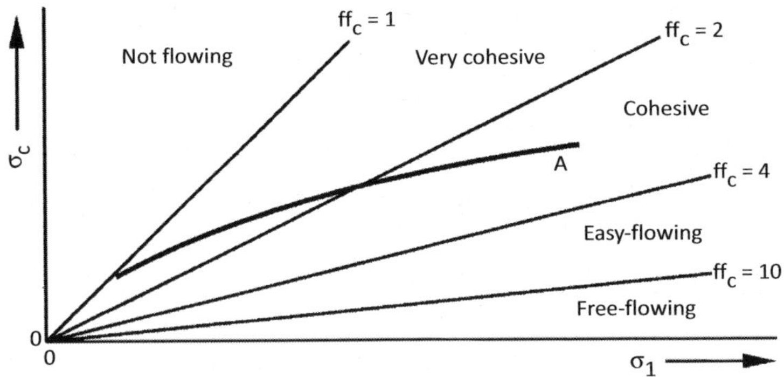

Figure 7   Flow function and lines of constant flowability.

of flowability on consolidation stress, it is not possible (unfortunately!) to describe the flowability of a bulk solid with only one numerical value.

With the results of time consolidation tests, flowability can be determined with Eq. (4), using the unconfined yield strength, σ_c, which was measured after the corresponding storage period. If the bulk solid shows a time consolidation effect, one will measure an increasing unconfined yield strength with increasing storage period, so that from Eq. (4) lesser flowability will follow. This is logical: If a bulk solid gains strength with an increasing period of storage at rest at a certain consolidation stress, it will be more difficult to get this bulk solid to flow; i.e., its flowability decreases with increasing storage period.

In Figure 5 flow function A and two time flow functions are shown. The flow function represents unconfined yield strength, σ_c, in dependence on consolidation stress, σ_1, without influence of a storage period, i.e., for the storage period $t = 0$. A time flow function represents the unconfined yield strength which emerges after storage at the consolidation stress over a period of time, t. The flow function and time flow functions from Figure 5 are shown in Figure 8.a along with the boundaries of the ranges, which follow from the classification of flowability as outlined above. It can be seen that flowabilities, $f\!f_c$, measured at identical consolidation stress, but after different consolidation periods, decrease with increasing consolidation time (Figure 8.b). For the consolidation stress $\sigma_{Example}$ chosen as an example, one obtains measurement points in areas of decreasing flowability when increasing the consolidation period t (see arrow in Figure 8.a).

From the dependence of flowability, $f\!f_c$, on consolidation stress, σ_1, it follows that one can compare the flow behaviour of several bulk solids quantitatively using $f\!f_c$ only if all measurements have been performed at

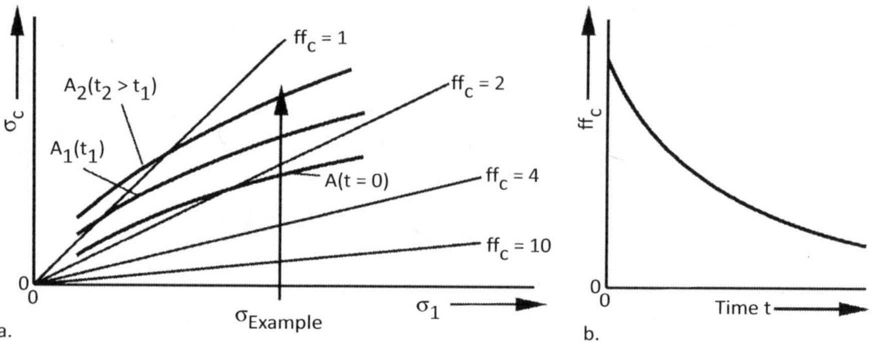

Figure 8   Influence of storage time on flowability.

identical consolidation stresses. Otherwise totally different (incorrect) statements might result. This shows how important it is to test a bulk solid at defined and known conditions (e.g., known consolidation stress).

5   Practical determination of flow properties

In the previous section the flow behaviour has been explained in a simplified way by using the uniaxial compression test as a model. The use of the uniaxial compression test with fine-grained, cohesive bulk solids is problematic, because one obtains unconfined yield strength values that are too low [7], and preparation of the hollow cylinder to obtain frictionless walls is very time-consuming. In addition, further important parameters (e.g., internal friction and wall friction) cannot be determined with this test. It is, however, an appropriate measurement technique for the measurement of the time consolidation of coarse-grained bulk solids.

In order to measure the flow properties of fine-grained bulk solids, in advanced bulk solids technology so-called shear testers are used. In the following first the principle of shear testing is outlined. Afterwards, the translational shear tester introduced by Jenike around 1960 (Jenike shear tester; the first shear tester especially designed for bulk solids) [1, 2, 5, 16] and the Schulze ring shear tester [2, 9–11, 14, 17] will be described.

5.1   Shear test procedure (yield locus)

The goal of a shear test is to measure the yield limit of a consolidated bulk solid. The yield limit is called yield locus in bulk solids technology. For a shear test, a bulk solid specimen is loaded vertically by a normal stress, σ (Figure 9.a). Then a shear deformation is applied on the specimen by moving the top platen with a constant velocity, v. This results in a horizontal shear stress, τ (Figure 9.b). With increasing shear stress the resultant force, F_R, acting on the bulk solid specimen, increases.

When a point of a yield locus is measured, in analogy to the uniaxial compression test, two steps are necessary: First the bulk solid specimen is consolidated, what is called "preshear." Subsequently a point of the yield limit is measured. This step is called "shear" or "shear to failure."

For preshear the bulk solid specimen is loaded in the vertical direction by a well-defined normal stress, $\sigma = \sigma_{pre}$. Then the specimen is sheared. At the beginning of preshear the shear stress τ increases with time (as shown in the left diagram in Figure 10). With time the curve of shear stress vs. time becomes flatter, and finally the shear stress remains constant even though the specimen is sheared further. The constant shear stress is called τ_{pre}.

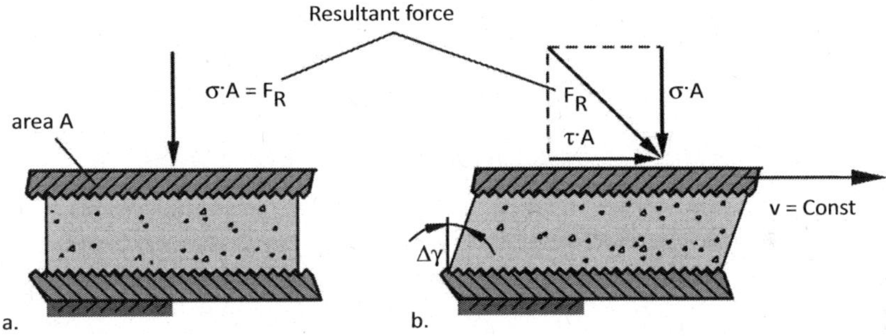

Figure 9 Bulk solid specimen: a. initial loading with normal stress σ; b. shear deformation [velocity v = const].

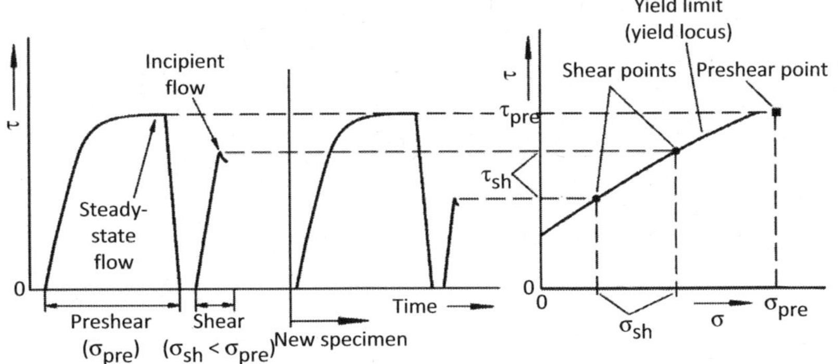

Figure 10 Plot of shear stress vs. time; yield locus.

After constant shear stress has been attained, neither shear resistance (and strength) nor bulk density increase further. Thus the bulk solid specimen is sheared at constant normal stress, σ, constant shear stress, τ, and constant bulk density, ρ_b. Thus flow, or plastic deformation, occurs at constant bulk density. This type of flow, attained at preshear, is called steady-state flow. The state of the bulk solid after steady-state flow is attained is called "critically consolidated with respect to normal stress, σ_{pre}." The characteristic stress for this consolidation—the major principal stress σ_1—will be considered later.

The bulk density, ρ_b, and the shear stress, τ_{pre}, attained at steady-state flow are characteristic for the applied normal stress at preshear, σ_{pre}. In principle, an identical state of consolidation, characterized by the same bulk density, ρ_b, and the same shear stress, τ_{pre}, will be attained with other specimens of the same bulk solid presheared under the same normal stress, σ_{pre}.

After the bulk solid specimen has been consolidated by the preshear procedure, the shear deformation is reversed until the shear stress, τ, is reduced to zero. The pair of values of normal stress and shear stress at steady-state flow $(\sigma_{pre}, \tau_{pre})$ is plotted in a normal stress-shear stress diagram (σ,τ-diagram, Figure 10, right). Point $(\sigma_{pre}, \tau_{pre})$ is called the "preshear point."

After preshear the bulk solid specimen in the shear cell is defined as a critically consolidated specimen. The second step of the test procedure—shear or shear to failure—is discussed next.

For shear to failure the normal stress acting on the specimen is decreased to a value σ_{sh}, which is less than the normal stress at preshear, σ_{pre}. Had the specimen been presheared under the lower normal stress, σ_{sh}, and not under σ_{pre}, its bulk density and strength would have been less. Since the specimen was presheared under the greater normal load, σ_{pre}, it was consolidated more than it would have been with the lower normal load, σ_{sh}.

If the consolidated specimen is sheared under the normal stress $\sigma_{sh} < \sigma_{pre}$, it will start to flow (fail) when a sufficiently large shear force, or shear stress, is attained. At that point particles start to move against each other. The material will start to dilate (decrease in bulk density) and shear resistance and thus shear stress will decrease (Figure 10). The maximum shear stress characterizes incipient flow. The corresponding pair of values (σ_{sh}, τ_{sh}) is a point of the yield limit of the consolidated specimen in the σ,τ-diagram (Figure 10, right). Such a point is called a "shear point" or a "point of incipient flow."

In order to measure the course of the yield locus, several of the tests described above must be performed, where the specimens first must be consolidated at identical normal stress, σ_{pre} (preshear). Then the specimens are sheared (to failure) under different normal stresses, $\sigma_{sh} < \sigma_{pre}$. As outlined above, by preshearing at identical normal stress, σ_{pre}, each specimen reaches the same state of consolidation. Each test yields the same preshear point (σ_{pre}, τ_{pre}), and one individual shear point (σ_{sh}, τ_{sh}) in accordance with the different normal stresses, σ_{sh}, applied at shear. The yield locus follows from a curve plotted through all measured shear points (Figure 10, right).

5.2   Jenike shear tester

Around 1960 Jenike [1] published his fundamental work on silo and bulk solids technology and introduced the Jenike shear tester, a translational shear tester. This tester was the first one designed for the purposes of powder technology (e.g., measurement at small stresses), and even today shear testers are compared to the Jenike shear tester.

The shear cell of the Jenike shear tester consists of a bottom ring (also called mould ring), a ring of the same diameter (so-called upper ring) lying above the bottom ring, and a lid (Figure 11). The lid is loaded centrally with a normal force, F_N. The upper part of the shear cell is displaced horizontally against the fixed bottom ring by a motor driven stem which pushes against a bracket fixed to the lid. The force F_S—the shear force—exerted by the stem is measured. Due to the displacement of the upper ring and the lid against the bottom ring, the bulk solid undergoes a shear deformation. The normal stress, σ, and shear stress, τ, acting in the horizontal plane between upper ring and

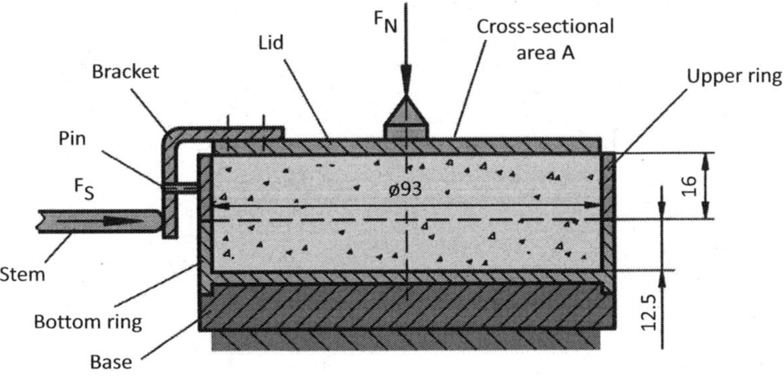

Figure 11   Shear cell of the Jenike shear tester [1,16].

bottom ring are determined by dividing normal force, F_N, and shear force, F_S, by the cross-sectional area of the shear cell, A.

For the measurement of a point of a yield locus, the shear cell is filled with the bulk solid specimen. After a manual preconsolidation [16] the specimen is presheared and then sheared to failure as outlined in the previous section. For the next point of the yield locus, a new bulk solid specimen has to be prepared and sheared.

Although the Jenike shear tester is internationally recognized, from today's point of view a disadvantage might be the time required for a test (one to two hours per yield locus; depending on the powder and the operator's skill) during which the operator has to be present. In addition, the manual preconsolidation of each specimen can be a source of measurement errors, and due to the limited shear displacement (maximum: twice the thickness of the wall of the upper ring) materials requiring too much deformation to attain steady-state flow can hardly be tested.

5.3   Ring shear tester

Ring shear testers (rotational shear testers) have been used in soil mechanics since the 1930s [4]. In the 1960s Walker designed a ring shear tester for bulk solids [3], where lower stresses than in soil mechanics are of interest. In the following decades different ring shear testers have been built and investigated at several universities (e.g., [6,8]). In 1992 a ring shear tester (type RST-01.01) [9,10] was developed by the author, followed by a computer-controlled version in 1997 (type RST-01.pc). It is connected to a personal computer running a control software. With this control software yield loci, wall yield loci, time consolidation, etc. can be measured automatically. A smaller computer-controlled ring shear tester (type RST-XS) has been available since 2002. This tester enables use of small specimen volumes (3.5 ml, 9 ml, 30 ml, and 70 ml).

Figure 12 shows the principle of the shear cell of a ring shear tester (series RST-01) [2,9,10,17]. The ring-shaped (annular) bottom ring of the shear cell contains the bulk solid specimen. The (annular) lid is placed on top of the bulk solid specimen. The lid is fixed at a crossbeam.

A normal force, F_N, is exerted to the crossbeam in the rotational axis of the shear cell and transmitted through the lid to the bulk solid specimen. Thus a normal stress σ is applied to the bulk solid specimen. The counterbalance force, F_A, also acts in the centre of the crossbeam. F_A is directed upward and

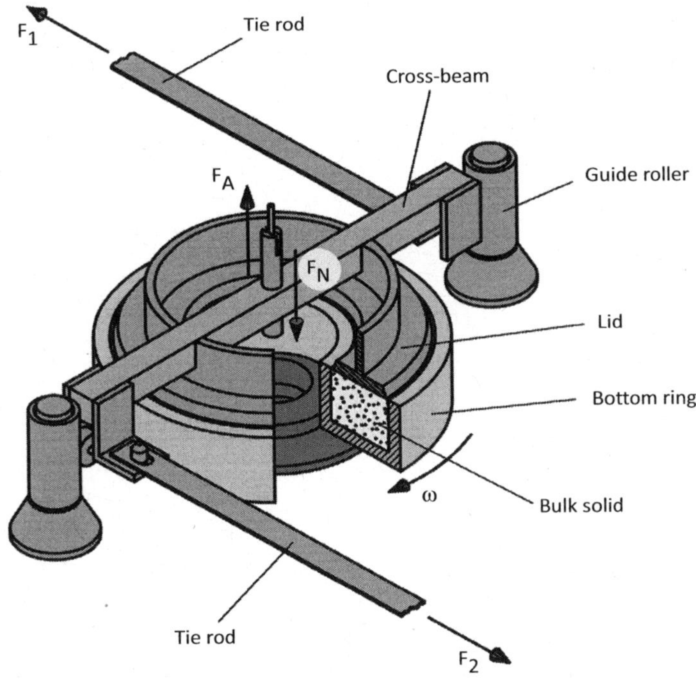

Figure 12 Shear cell of a ring shear tester (here: Schulze ring shear tester type RST-01) [2,9,10,17].

is created by counterweights. F_A counteracts the gravity forces of the lid, the hanger, and the crossbeam.

To shear the bulk solid, the lid and the bottom ring of the shear cell must rotate relative to each other. This is accomplished by rotating the bottom ring in the direction of the arrow ω (ω is the angular velocity), whereas the lid and the crossbeam are prevented from rotation by two tie rods connected to the crossbeam. Each of the tie-rods is fixed at a load beam, so that the forces, F_1 and F_2, acting in the tie rods can be measured.

The bottom of the shear cell and the lower side of the lid are rough in order to prevent the bulk solid from sliding relative to these surfaces. Therefore, rotation of the bottom ring relative to the lid creates a shear deformation within the bulk solid. Through this shearing the bulk solid is deformed and thus a shear stress τ prevails. The forces acting in the tie rods (F_1 and F_2) are directly proportional to the shear stress τ acting in the bulk solid.

In addition to the shear force (F_1, F_2), the ring shear tester also measures the vertical position of the lid. If a bulk solid is compressible, its bulk density,

ρ_b, will increase more or less, depending on the normal load applied. If the vertical position of the lid relative to the bottom ring is measured, one can calculate the volume of the bulk solid specimen. If the mass of the specimen is also determined through weighing, bulk density can be calculated.

The principle of the ring shear tester described above is different from other ring shear testers in a few important details, e.g., the lid, which lies on the bulk solid sample, similar to the lid of the Jenike shear cell, and is not held horizontally by a bearing as with older ring shear testers. This yields a more homogeneous stress distribution across the specimen, and the absence of any bearing friction increases the accuracy of the measurement. Additionally, the masses of the lid and all parts connected to it are small, so that tests at low normal stresses are possible, and the shear cell including the lid and the bulk solid specimen can be taken from the tester without disturbing the sample, e.g., for time consolidation tests using a time consolidation bench.

The test procedure (Figure 13) is quite similar to the test procedure recommended for the Jenike tester (preshear and shear, see above), although the test procedure for the ring shear tester is less time consuming, easier to perform, and, hence, less influenced by the person who runs the test. This in combination with the design outlined in the last paragraph results in the good reproducibility compared to other testers [18].

With the ring shear tester usually a complete yield locus is measured with one specimen (in contrast to the Jenike tester where only one point can be measured with one specimen). In order to measure another point of the yield

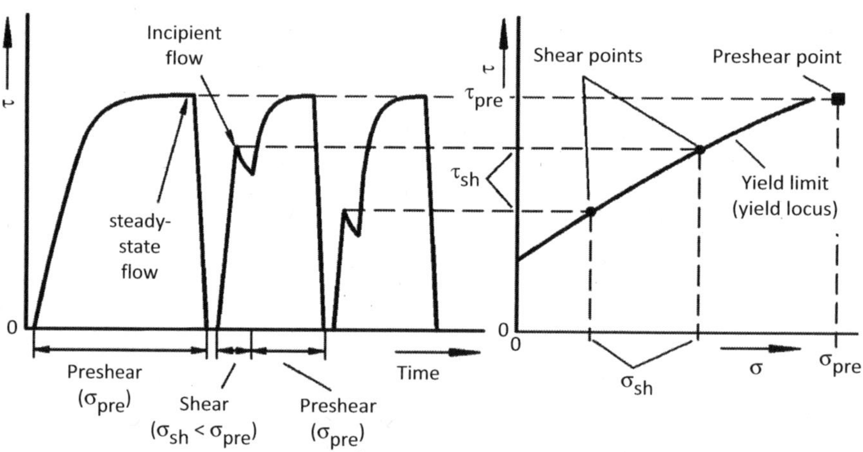

Figure 13   Shear test procedure of a ring shear tester.

limit with the same specimen, after shear (shear to failure) the normal stress is increased again to σ_{pre}, which is the normal stress applied at first preshear. Then the specimen is presheared again under this normal stress until steady-state flow is attained. Thus, the specimen is again critically consolidated. After the specimen is relieved from shear stress (backward rotation of the shear cell until $\tau = 0$), the normal stress is reduced to another value of $\sigma_{sh} < \sigma_{pre}$ (Figure 13), and the specimen is then sheared again, thus obtaining another point of the yield limit in the σ,τ-diagram. After shear, the specimen is again presheared, then sheared, and so on, until a sufficient number of points of the yield limit are known and the yield locus can be drawn.

5.4 Yield locus

The parameters which describe the flow properties can be determined from the yield locus (Figure 14). The relevant consolidation stress σ_1 is equal to the major principal stress of the Mohr stress circle, which is tangential to the yield locus and intersects at the point of steady state flow (σ_{pre}, τ_{pre}). This stress circle represents the stresses in the sample at the end of the consolidation procedure (stresses at steady state flow). It corresponds to the stress circle at the end of consolidation at the uniaxial compression test (Figure 3). The unconfined yield strength, σ_c, results from the stress circle which is tangential to the yield locus and which runs through the origin (minor principal stress $\sigma_2 = 0$). This stress circle represents a similar stress state as the one which prevails in the second step of the uniaxial compression test (stress circle B_3, Figure 4). In contrast to the uniaxial compression test the unconfined yield strength, σ_c, has to be determined on basis of the yield locus and does not follow directly from the measurement.

Please note that the analogy between the uniaxial compression test and the shear test is used here for the explanation of the yield locus. In reality, the stress circles at uniaxial compression and at steady state flow are not exactly the same, and a uniaxial compression test usually results in a smaller unconfined yield strength than a shear test [2, 7, 12, 13, 20].

A straight line through the origin of the σ,τ-diagram, tangent to the greater Mohr circle (steady-state flow), is the effective yield locus as defined by Jenike [1] (broken line in Figure 14). It encloses the σ-axis with the angle ϕ_e (effective angle of internal friction). Because the largest Mohr stress circle indicates a state of steady-state flow, the angle ϕ_e can be regarded as a

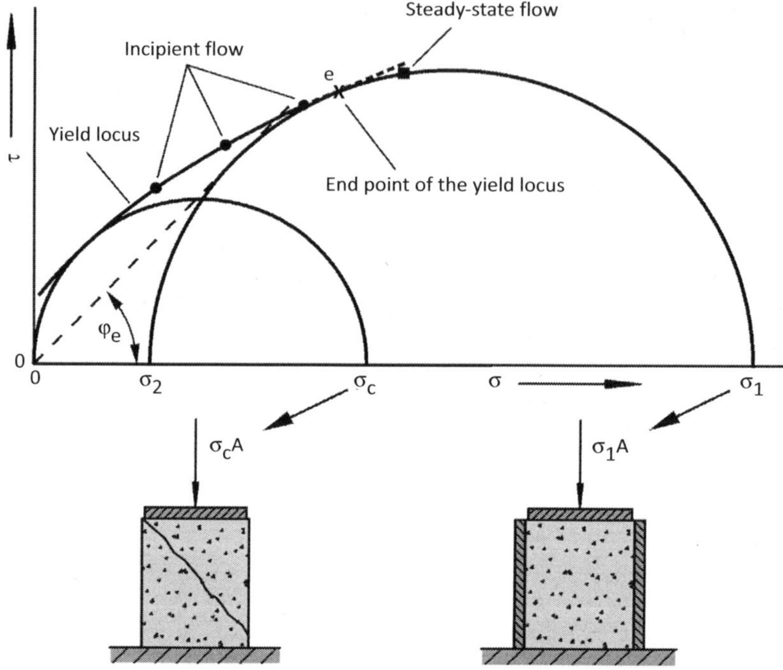

Figure 14   Yield locus, analogy to uniaxial compression test.

measure of the internal friction at steady-state flow. This angle is required for silo design according to Jenike's theory.

If several yield loci are measured at different stress levels, i.e., with different normal stresses at preshear, σ_{pre}, each yield locus represents another state of consolidation and another bulk density. The above-mentioned flow properties (unconfined yield strength, effective angle of internal friction) can be indicated as a function of the consolidation stress, σ_1, similar to Figure 4 where bulk density and unconfined yield strength are plotted vs. the consolidation stress.

5.5   Time consolidation

The effect of time consolidation has been outlined in Section 4.2. Before time consolidation can be measured, a yield locus must be determined at the same consolidation stress.

The time consolidation, which describes the increase of the unconfined yield strength with time during storage at rest, is measured with a shear tester similar to the measurement of a yield locus. First a bulk solid specimen is presheared (consolidated). After preshear the specimen is stored for a period, t, under the vertically acting normal stress, σ, which is selected to be equal to the consolidation stress, σ_1, of the corresponding yield locus. This ensures that during the consolidation period the same major principal stress (= consolidation stress, σ_1) acts on the specimen as during steady-state flow at preshear.

After the time consolidation period t the specimen is sheared to failure. For this, a vertical normal load, $\sigma_{sh} < \sigma_1$, is selected. As with shear without time consolidation (measurement of a point of a yield locus), so also after time consolidation will one observe a shear stress maximum. If consolidation time affects the bulk solid under consideration, after the consolidation period the shear stress maximum will be larger than it would have been without a consolidation period between preshear and shear (Figure 16).

The maximum shear stress, τ, is a point of a yield limit, which is valid for the applied storage period, t, and called a "time yield locus." Figure 17 shows a yield locus and two time yield loci obtained for different consolidation periods, t_1 and t_2. The yield locus can also be regarded as a time yield locus for $t = 0$.

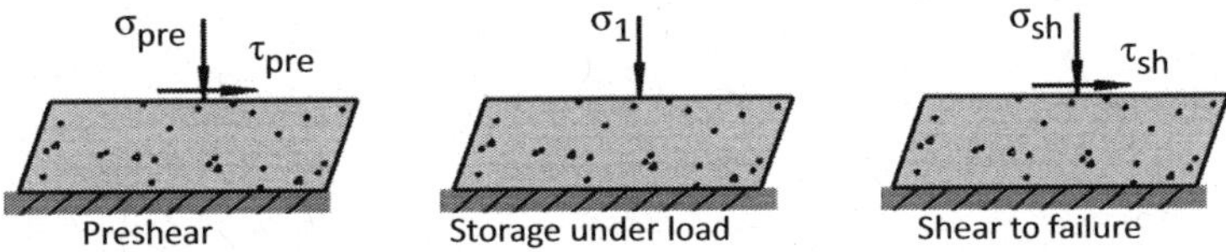

Figure 15   Time consolidation test.

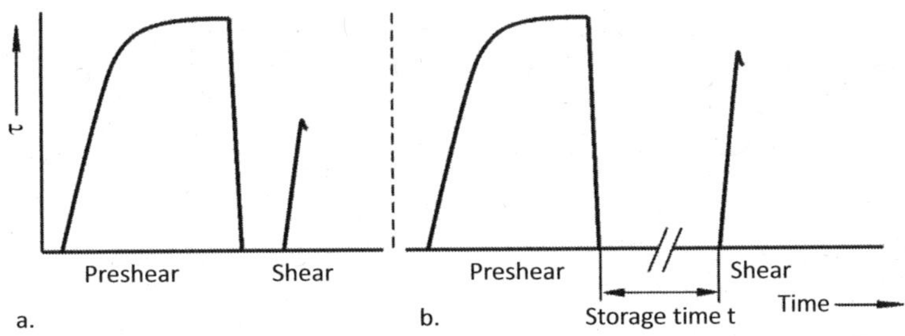

Figure 16   Shear stress vs. time at shear: with (a) and without (b) time consolidation.

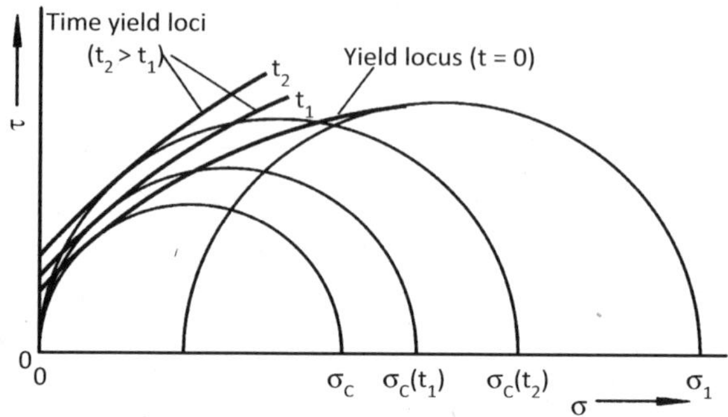

Figure 17   Yield locus and time yield loci.

With the measured shear points a time yield locus can be approximated similarly to the approximation of a yield locus. Compared to the yield locus, the time yield locus is shifted toward greater shear stresses, τ (if the bulk solid shows an increase of strength with time). The unconfined yield strength, σ_c, is determined in the same way as for a yield locus by drawing a Mohr stress circle through the origin and tangent to the time yield locus. In Figure 17 the values of the unconfined yield strength for the consolidation periods, t_1 and t_2, are designated as $\sigma_c(t_1)$ and $\sigma_c(t_2)$.

Time yield loci can be determined for different storage periods (consolidation periods). Each time yield locus is valid for only one consolidation period and one consolidation stress. If the strength of the bulk solid increases over time, the time yield loci will be shifted toward larger values of τ as the consolidation period, t, increases (see Figure 17, $t_2 > t_1$)

5.6   Wall friction

Wall friction is the friction between a bulk solid and the surface of a solid, e.g., the wall of a silo or a bin. The coefficient of wall friction or the wall friction angle, respectively, is important both for silo design for flow and silo design for strength, but also for the design of chutes and other equipment, where the bulk solid will flow across a solid surface. Knowing the wall friction angle, it is possible to decide whether or not the polishing of the wall surface or the use of a liner would have advantages in the flow of the bulk solid.

The *principle* of a wall friction test, where the kinematic angle of wall friction is determined, is shown in Figure 18. The bulk solid specimen is

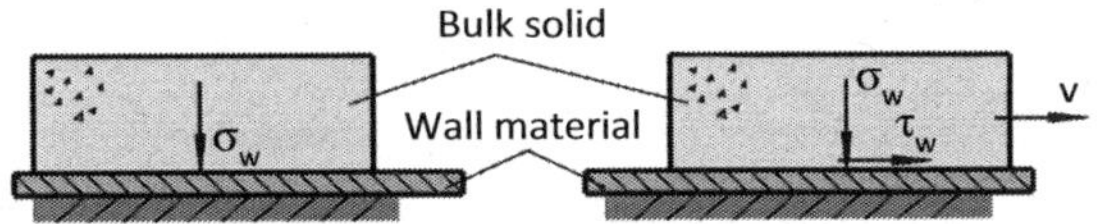

Figure 18   Principle of a wall friction test.

subjected to a vertical normal stress. The normal stress acting between bulk solid specimen and wall material is called the wall normal stress, σ_w.

The bulk *solid* specimen is then shifted relative to the wall material surface with a constant velocity, v. This process is called shear (similar to the yield locus test). The shear stress acting between bulk solid specimen and wall material is measured.

It is *usual* to measure wall friction at incrementally decreasing wall normal stresses [1,16]. Thus one begins with the greatest wall normal stress (σ_{w1} in Figure 19). At the beginning of the shear process the wall shear stress, τ_w, increases. With time, the increase of the wall shear stress becomes less until finally a constant wall shear stress, τ_{w1}, is attained (steady-state shear stress). The constant wall shear stress, τ_{w1}, is characteristic for the applied wall normal stress, σ_{w1}. After the steady-state condition is attained, the normal load is reduced. With each decrease in wall normal stress, wall shear stress, τ_w, also decreases (Figure 19). After a certain time, a steady-state shear stress is again attained. In this way values of steady-state wall friction at several wall normal stresses are measured.

The pair of values of wall normal stress and constant wall shear stress (σ_w, τ_w) describes the kinematic wall friction at the wall normal stress, σ_w, and is used for the evaluation of the test. All pairs of values of wall normal stress and steady-state wall shear stress are plotted in a σ_w,τ_w-diagram (Figure 19, right). The curve (or line) running through the measured points is called the wall yield locus.

The wall yield locus is a yield limit like the yield locus. The wall yield locus describes the wall shear stress, τ_w, necessary to shift a bulk solid continuously across a wall surface under a certain wall normal stress, σ_w. Since the wall yield locus is based on the shear stresses measured at steady-state conditions, it describes the kinematic friction of the bulk solid. Thus the wall yield locus could more exactly be called a kinematic wall yield locus [16].

To quantify wall friction, the wall friction angle, ϕ_x, or the coefficient of wall friction, μ, are used. The larger the wall friction angle or coefficient of wall friction, the greater is wall friction. The coefficient of wall friction, μ, is the ratio of wall shear stress, τ_w, to wall normal stress, σ_w. The wall friction

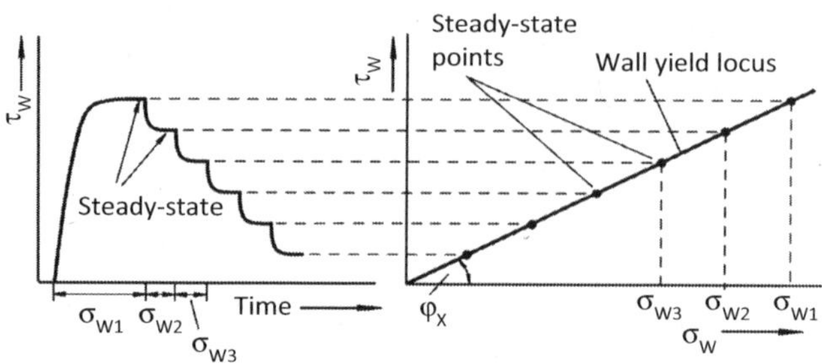

Figure 19   Course of wall shear stress in a wall friction test; wall yield locus.

angle, ϕ_x, is the slope of a line running through the origin of the σ_w,τ_w-diagram and a point of the wall yield locus.

If the wall yield locus is a straight line running through the origin (Figure 19), the ratio of wall shear stress, τ_w, to wall normal stress, σ_w, has the same value for each point of the wall yield locus. Thus one obtains the identical wall friction coefficient, μ, and the identical wall friction angle, ϕ_x, for each point of the wall yield locus. In this case wall friction is independent of wall normal stress.

The wall yield locus shown in Figure 20 is curved and does not run through the origin. In this case one finds a different wall friction coefficient and wall friction angle for each point of the wall yield locus. Thus the wall friction coefficient and the wall friction angle are dependent on wall normal stress, σ_w. This can be seen by the wall friction angles, ϕ_{x1} and ϕ_{x2}, which follow for the wall normal stresses, σ_{w1} and σ_{w2}. A wall yield locus intersecting the τ-axis at $\tau_{ad} > 0$ is typical for materials tending to adhere to walls (e.g., like moist clay). The shear stress, τ_{ad}, at the point of intersection is called adhesion.

Wall friction can be measured with the shear testers described above. The setup of the Jenike shear tester for a wall friction test is shown in Figure 21. The bottom ring of the shear cell is replaced by a sample of wall material (e.g., stainless steel, coated steel). The wall normal stress is then adjusted by the normal force, F_N, and the shear force, F_S, is measured following the procedure outlined above.

Figure 22 shows the setup of the wall friction shear cell of the Schulze ring shear tester [2,9,10,17]. The annular bottom ring contains the sample of the wall material. On top of the wall material sample is the bulk solid specimen, which is covered with the annular lid of the shear cell. The lid is connected to the crossbeam. Except for the geometry of bottom ring and lid,

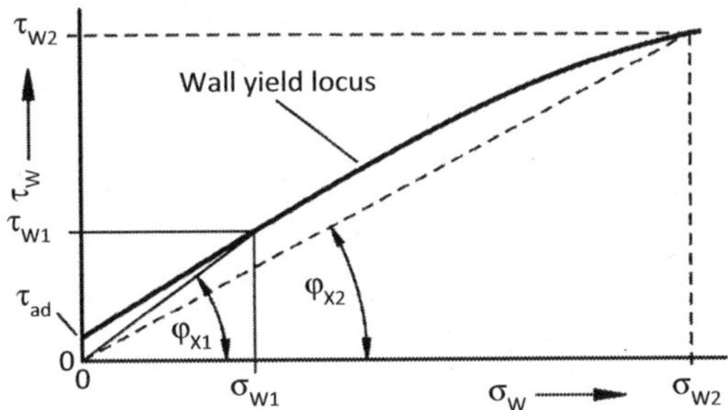

Figure 20   Wall yield locus; wall friction angle is dependent on the wall normal stress.

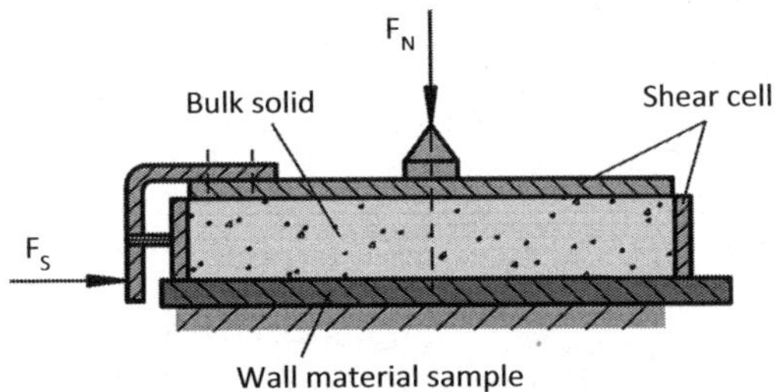

Figure 21   Measurement of wall friction with the Jenike shear tester [1].

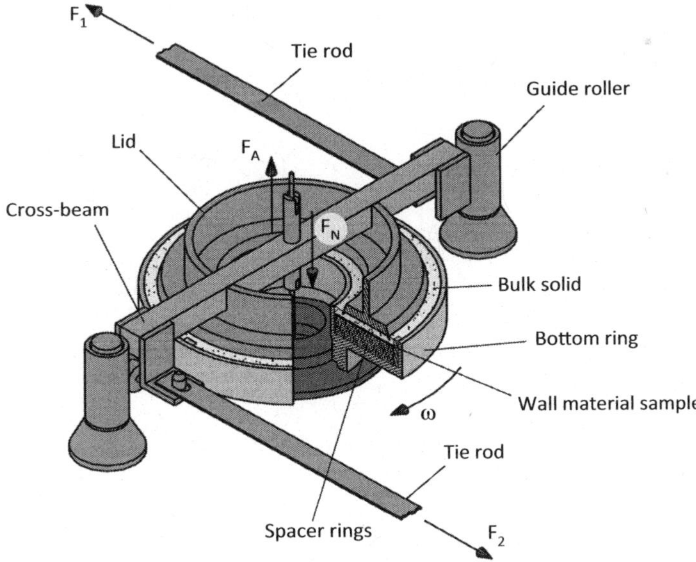

Figure 22   Setup of the shear cell for wall friction tests (Schulze ring shear tester) [2, 9, 10, 17].

the setup is similar to the setup of the shear cell for flow properties testing (Section 5.3).

To measure wall friction the shear cell is rotated slowly in the direction of arrow ω, while the lid is prevented from rotating by the two tie rods. The forces acting on the tie rods, F_1 and F_2, are measured. The layer of bulk solid located between the lid and the surface of the wall material sample is prevented from rotating by the lid, which has a rough underside. Thus the bulk solid is shifted across the surface of the wall material sample while it is subjected to the normal stress, σ_w. The wall shear stress, τ_w, is calculated from the F_1 and F_2.

6   Further measurement methods and devices for the determination of bulk solid properties

Several empirical methods are used for the assessment of bulk solid properties, e.g., the determination of the angle of repose, α_M (Figure 23). Different results are obtained even with this simple test procedure: So a conical heap (a) will yield a different angle of repose α_M compared to a wedge-shaped heap. If the bulk solid flows out of a container with a central outlet (b), then the angle of repose α_M will be even higher. In a rotating drum another (smaller) angle of repose will prevail.

Examples of further simple test methods are (overview see [2, 12, 13, 20, 21]):

- Determination of the time required for a sample of bulk solid to flow out of a model silo.

- Determination of the compressibility by measuring loose density and tap density (Hausner ratio).

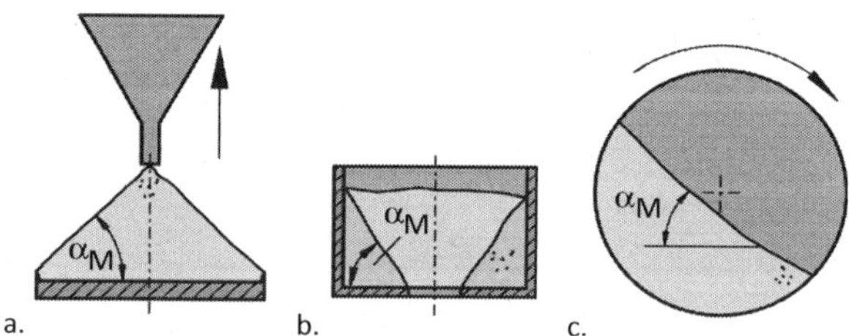

Figure 23   Methods to measure the angle of repose.

- The Carr index determined from compressibility, different angles of repose, and a particle size distribution [19].

The results of these so-called simple tests depend on particular boundary conditions (e.g., in the case of determining the angle of repose, the conditions at the free surface of the heap) which give no information about the behaviour of a bulk solid under the stresses which are present in a typical application, e.g., in a container or in a hopper.

- The results depend on the test devices; therefore the physical figures are not device-independent.

- The influences which are important for the design of silos (storage time, stress level) are not included or only measured qualitatively.

- Many of the simple test methods cannot reasonably be applied to cohesive bulk solids with poor flowability (flour, zinc oxide).

- The preparation of the bulk solid influences the measured results (e.g., the thorough stirring of bulk solids which easily fluidise leads to smaller angles of repose).

Because of the weaknesses in these simple test methods mentioned above, test devices which allow the defined preparation of the sample of bulk solid must be preferred (e.g., preshear until the onset of steady state flow as a defined consolidation procedure). Only in this way it is possible to determine the relevant flow properties (unconfined yield strength, internal friction angle, wall friction angle, bulk density, time consolidation dependent on the consolidation stress) independent from the devices used [12, 13].

7 Summary

Consolidated bulk solids have yield limits like other materials. These yield limits, which can be measured with shear testers, are called yield loci. They depend on the consolidation which has previously taken place and, sometimes, also on the storage time. If a bulk solid is to be set in motion, e.g., to flow out of a silo, the stresses acting on the bulk solid must be large enough to ensure that the corresponding Mohr stress circle touches the yield locus of the consolidated bulk solid.

The flow properties which can be obtained from the measured yield loci are exactly defined physical figures. Unconfined yield strength (dependent on consolidation stress and storage time), bulk density, and wall friction angle are the most important flow properties for the design of storage and conveying systems as well as for comparative tests, quality control, and product development.

Literature

[1] Jenike, A.W.: Storage and flow of solids, Bull. No. 123, Engng. Exp. Station, Univ. Utah, Salt Lake City (1964)

[2] Schulze, D.: Powders and Bulk Solids – Behavior, Characterization, Storage and Flow. Springer, Berlin – Heidelberg – New York – Tokyo (2008)

[3] Carr, J.F., Walker, D.M.: An annular shear cell for granular materials, Powder Technology 1 (1967/68), pp. 369–373

[4] Hvorslev, M.J.: Über die Festigkeitseigenschaften gestörter bindiger Böden, Ingeniørvidenskabelige Skrifter A, Nr. 45 (1939)

[5] Molerus, O.: Schüttgutmechanik. Springer Verlag, Berlin – Heidelberg – New York – Tokyo (1985)

[6] Gebhard, H.: Scherversuche an leicht verdichteten Schüttgütern unter besonderer Berück sichtigung des Verformungsverhaltens, Diss. Univ. Karlsruhe (1982)

[7] Schwedes, J., Schulze, D.: Measurement of flow properties of bulk solids, Powder Technology 61 (1990), pp. 59–68

[8] Münz, G.: Entwicklung eines Ringschergerätes zur Messung der Fließeigenschaften von Schüttgütern und Bestimmung des Einflusses der Teilchengrößenverteilung auf die Fließeigenschaften kohäsiver Kalksteinpulver, Diss. Univ. Karlsruhe (1976)

[9] Schulze, D.: Development and application of a novel ring shear tester, Aufbereitungstechnik 35 (1994) (10), pp. 524–535

[10] Schulze, D.: A new ring shear tester for flowability and time consolidation measurements, Proc. 1st International Particle Technology Forum, August 1994, Denver, Colorado, USA, pp. 11–16

[11] Schulze, D.: Appropriate devices for the measurement of flow properties for silo design and quality control, PARTEC 95, Preprints "3rd Europ. Symp. Storage and Flow of Particulate Solids", 21–23.3.95, Nürnberg, pp. 45–56

[12] Schulze, D.: Zur Fließfähigkeit von Schüttgütern – Definition und Meßverfahren, Chem.-Ing.-Techn. 67 (1995) (1), pp. 60–68

[13] Schulze, D.: Flowability of bulk solids – Definition and measuring techniques, Part I and II, Powder and Bulk Engineering 10 (1996) (4), pp. 45–61, and 10 (1996) (6), pp. 17–28

[14] Schulze, D.: Flowability and time consolidation measurements using a ring shear tester, Powder Handling & Processing 8 (1996) (3), pp. 221–226

[15] Kwade, A., Schulze, D, Schwedes, J.: Determination of the stress ratio in uniaxial compression tests, Powder Handling & Processing 6 (1994) (1), pp. 199–203

[16] The Institution of Chemical Engineers (Eds.): Standard shear testing technique for particulate solids using the Jenike shear cell (1989)

[17] ASTM Standard D6773-02: Standard shear test method for bulk solids using the Schulze ring shear tester, ASTM International, www.astm.org

[18] Verlinden, A.: Experimental assessment of shear testers for measuring flow properties of bulk solids, PhD-Thesis, Univ. of Bradford, UK (2000)

[19] Carr, R.L.: Classifying flow properties of solids, Chem. Engng. 72 (1965) (3), pp. 69–72

[20] Schulze, D.: The measurement of the flowability of bulk solids. In: Brown CJ, Nielsen J (Eds.) Silos – Fundamentals of theory, behaviour and design. E & FN Spon, London and New York (1998), pp. 18–52

[21] Svarovski, L.: Powder testing guide. Elsevier Applied Science Publishers Ltd., London and New York (1987)

Index

Other Manufacturing Titles Certain to Be of Interest

- Advanced Regulatory Control: Applications and Techniques, David W. Spitzer
- Alarm Management for Process Control, Douglas H. Rothenberg, Ph.D
- Centrifugal and Axial Compressor Control, Gregory K. McMillan
- Industrial Resource Utilization and Productivity, Edited by Anil Mital, Ph.D. and Arun Pennathur, Ph.D.
- Protecting Industrial Control Systems from Electronic Threats, Joe Weiss
- Process Control Case Histories: An Insightful and Humorous Perspective from the Control Room, Edited by Gregory K. McMillan
- Robust Control System Networks: How to Achieve Reliable Control After Stuxnet, Ralph Langner
- WBF: Applying ISA-88 in Discrete and Continuous Manufacturing
- WBF: ISA-88 and ISA-95 in the Life Science Industries,
- WBF: ISA-95 Implementation Experiences
- WBF: ISA-88 Implementation Experiences
- Social Media for Engineers and Scientists, Jon DiPietro
- The Engineering Language: A Consolidation of the Words and Their Definitions, Ronald Hanifan
- Reduce Your Engineering Drawing Errors, Ronald Hanifan

For more information, please visit www.momentumpress.net

The Momentum Press Digital Library
Engineering Ebooks For Research, Classrooms, and Reference

Our books can also be purchased in an e-book collection that features…

- *a one-time purchase that is owned forever; not subscription-based*
- *allows for simultaneous readers,*
- *has no restrictions on printing*
- *can be downloaded as a pdf*

The Momentum Press digital library is an affordable way to give many readers simultaneous access to expert content.

For more information, please visit www.momentumpress.net/library
and to set up a trial, please contact mptrials@globalepress.com.